Travels in Four Dimensions

Travels in Four Dimensions

The Enigmas of Space and Time

Robin Le Poidevin

OXFORD
UNIVERSITY PRESS

OXFORD
UNIVERSITY PRESS

Great Clarendon Street, Oxford OX2 6DP

Oxford University Press is a department of the University of Oxford.
It furthers the University's objective of excellence in research, scholarship,
and education by publishing worldwide in

Oxford New York

Auckland Bangkok Buenos Aires Cape Town Chennai
Dar es Salaam Delhi Hong Kong Istanbul Karachi Kolkata
Kuala Lumpur Madrid Melbourne Mexico City Mumbai Nairobi
São Paulo Shanghai Taipei Tokyo Toronto

Oxford is a registered trade mark of Oxford University Press
in the UK and in certain other countries

Published in the United States
by Oxford University Press Inc., New York

British Library Cataloguing in Publication Data

Data available

Library of Congress Cataloging in Publication Data

Le Poidevin, Robin, 1962–
What are space and time?: a philosophical introduction / Robin Le Poidevin.
p. cm.
Includes bibliographical references (p.) and index.
1. Space and time. I. Title.
BD632 .L46 2003   115—dc21   2002035583

ISBN 0–19–875254–7 (hbk.)
ISBN 0–19–875255–5 (pbk.)

1 3 5 7 9 10 8 6 4 2

Typeset by Hope Services (Abingdon) Ltd.
Printed in Great Britain
on acid-free paper by
Biddles Ltd,
Guildford & King's Lynn

For my students at Leeds,
   past and present

What, then, are space and time? Are they real existences? Are they only determinations or relations of things, yet such as would belong to things even if they were not intuited? Or are space and time such that they belong only to the form of intuition, and therefore to the subjective constitution of our mind, apart from which they could not be ascribed to anything whatsoever?

Immanuel Kant, *Critique of Pure Reason*

All the vital problems of philosophy depend for their solutions on the solution of the problem what Space and Time are and more particularly how they are related to each other.

Samuel Alexander, *Space, Time and Deity*

PREFACE

In the house where I grew up was a copy belonging to my mother of the seven-volume *Newne's Pictorial Knowledge*, an encyclopaedia for children published in the early 1930s. As a child I would spend hours with these books. There was an item in the back of each volume that held a particular fascination for me, a series of leaves one could lift up thus revealing the internal structure of an oyster, a frog, a dog rose, a bee (especially frightening), and various other things. In one of the volumes, containing an account of famous scientists and inventors, was a rather improbable story of the friar Roger Bacon. Bacon, the story went, had after many years' labour constructed a marvellous human head in brass, which, he said, would presently speak of wonderful things. Tired of watching over the head and waiting for it to speak, he set another friar to guard it, who was instructed to fetch Bacon the moment the head spoke. After some considerable time, the lips of the brass head began to move, and spoke the words 'Time is'. Not thinking this sufficiently significant to fetch Bacon, the friar waited to see what else the head would say. After half an hour, the head spoke once more: 'Time was'. Again, the friar sat still. After another half-hour, the head spoke for a third and final time: 'Time is past'. It then dashed itself to pieces on the floor. The friar went at once to tell Bacon of the calamity. Bacon, dismayed to find that the head had spoken in his absence and was now destroyed, made many other heads in brass, but none of them spoke.

The contributor of this part of the encyclopaedia wisely warned its readers that the story was a legend, but even if not true, showed what esteem Bacon was held in by his contemporaries and succeeding generations. I happily ignored this warning, captivated as I was by a story I felt quite sure was true. It convinced me that there were mysteries of time that held the key to life, but also that knowledge of these mysteries was possibly dangerous, even forbidden, for human minds. So began my fascination with time, although I had little idea where to turn for enlightenment. My interest was rekindled a few years later when my father happened to mention to me, out of the blue, J. W. Dunne's *Experiment with Time,* a book that had enjoyed enormous popularity on its first publication in 1927, and which had influenced the time plays of J. B. Priestley. It was, my father explained, something to do with a dream about the face of a clock and appeared to demonstrate that one could see into the future, but had since been disproved. (Dunne's account of this dream appears at the end of this book.) I am not quite sure what disproof my father had in mind, but when, some time afterwards, I found a copy of the book, I was both excited by the accounts of dreams, but also disappointed by my inability to understand Dunne's theory to explain them—a theory that still strikes me as decidedly peculiar.

It was not until I became a research student that I began thinking about the philosophy of time, and I recall the extraordinary impact that McTaggart's proof of the unreality of time had on me when I first encountered it. It convinced me, first, that there was in reality no absolute distinction between past, present, and future, and secondly, that in consequence our view of ourselves as observers moving through time was radically mistaken. The intimate connection between time and the self is, surely, one of the sources of the fascination the philosophical paradoxes discussed in this book have for us. A great many other philosophical problems are affected by views on space and time, and I believe it is no exaggeration to say that these two lie at the heart of metaphysical inquiry.

This book arose out of a course of lectures I have given for many years at Leeds, entitled *Space, Time and Infinity.* My purpose in writing it was primarily to introduce the reader to the classic paradoxes and problems of space and time, where our philosophical thinking about these two elusive ideas begins. Introducing theories was very much a secondary aim. Although I offer in the pages that follow some theoretical apparatus, which I believe is helpful in structuring our first thoughts about the problems, I have tried to keep this fairly light. It is the problems themselves that stimulate independent thought, and my aim will be fulfilled if the reader is as excited about the problems as I have been and feels just as keenly the need to seek solutions to them. I have sketched some possible solutions, but I am no evangelist, and would encourage anyone to treat the results of my attempts with deep suspicion. To further stimulate independent thought, I have added some questions for the reader at the end of each chapter, and a set of problems at the end of the book. Most chapters are more discursive and open-ended than would be tolerable in a journal article, but where I think a certain line of thought is mistaken, I say so. And where I have a particular angle on a debate, I pursue it. Since the book begins with Kant's question 'What are space and time?', the reader naturally expects some kind of answer. The Concluding Thoughts section, however, should be taken as a summary of some of the key ideas expressed in the book, rather than a definite conclusion, which would not be appropriate in so introductory a volume. Those looking for a more thorough grounding in theory, less elementary introductions to the subject, or less compromising defences of a particular position, will find some suggestions in the section on Further Reading at the end of the book.

This is, let me emphasize, *a philosophical* introduction to space and time, one that is concerned throughout with the conceptual questions and difficulties that our ordinary views of space and time throw up. I have had to introduce a very modest amount of physics, as it is difficult to get far in discussing these problems without some reference to physics, but this is emphatically *not* a popular science book, nor is it an introduction to the philosophy of space-time physics. I do not, for example, discuss the Special or General Theories of Relativity. I regard conceptual analysis of the classic paradoxes and problems as an important preliminary to thinking about space-time physics. Again, anyone looking for books on the philosophy of space-time physics will find suggestions in Further Reading, but I should particularly like to mention in this context Barry Dainton's excellent *Time and Space,* which appeared just as this book was being completed. Dainton's book also takes some of the issues discussed here several stages further.

Those familiar with the literature will be only too aware of the influences on my approach. I gratefully acknowledge those whose writings have provided particularly important influences and sources of inspiration: Bas van Fraassen, Graham Nerlich, Bill Newton-Smith, Hugh Mellor, Huw Price, and Richard Sorabji.

Conversations with many friends and colleagues have affected my thinking about space, time, and related issues, and I would like to thank James Bradley (from whom I first learned about the Greenwich incident), Jeremy Butterfield, Peter Clark, Heather Dyke, Steven French, Jonathan Lowe, Hugh Mellor (to whom my greatest debt is owed), Mark Nelson, Sharon Ney, Nathan Oaklander, Peter Simons, Quentin Smith, and the late Murray MacBeath.

An unfailing source of inspiration over the years has been provided by the undergraduate and postgraduate students it has been my privilege to teach at Leeds. Tutorials spent discussing space and time were lively and enjoyable occasions, and a number of times I found myself with my back to the wall trying to defend some position that met with vigorous opposition. There are too many people to whom I am indebted for it to be possible to give an exhaustive list, but I would particularly like to mention Andrew Bennett, Catherine Cour-Palais (née Sale), Claudia Courtis, Louisa Dale, Jim Eccles, Nikk Effingham, Nathan Emmerich, Heather Fotheringham, Martin Gough, Nick Jones, Danni Lamb, Kathryn Le Grice (née Davies), Olaus McLeod, Danielle Matthews, Stephen Mumford, Rebecca Roache, Jason Sender, Chris Taylor, Alice Temple, Sean Walton, Tom Williams. To these, and to all my present and former students, this book is dedicated with gratitude and affection.

ACKNOWLEDGEMENTS

I should particularly like to record my thanks here to Peter Momtchiloff, of Oxford University Press, who first suggested that I write this book, provided encouragement when the project turned out to be a rather longer one than I had anticipated, and went well beyond the call of editorial duty in reading through the entire typescript and making numerous helpful suggestions.

I have been greatly helped at many points, and particularly in writing the final chapter, by Huw Price, who also read the entire typescript and provided much penetrating and insightful criticism and advice. My grateful thanks to him, both for this, and for drawing my attention to the *Rambler* letter referred to in Chapter 10.

Chapter 9 draws on my 'Zeno's Arrow and the Significance of the Present', in Craig Callendar (ed.), *Time, Reality and Experience* (Cambridge University Press, 2002). I am grateful to the editor for permission to reuse some of the material of that article.

Robin Le Poidevin

May 2002

CONTENTS

1. The Measure of All Things   I

    Incident at Greenwich   I

    Metric, Convention, and Fact   5

    Time and the Laws of Nature   8

2. Change   13

    Time as Change   13

    Time without Change?   17

    Everything Has a Reason   24

3. A Box with No Sides?   30

    Where Two Worlds Meet   30

    Aristotle against the Void   31

    Jars, Pumps, and Barometers   34

    Lessons of the Vacuum   36

    The Redundancy of Space   41

    The Search for Absolute Motion   44

4. **Curves and Dimensions** 52

   Euclid Displaced 52

   Space Makes Its Presence Felt 57

   The Lone Hand 62

   More than Three Dimensions? 66

5. **The Beginning and End of Time** 73

   Echoes of Creation, Portents of Armageddon 73

   The Limits of Reason 76

   Can the Past be Infinite? 79

   The Great Circle 83

6. **The Edge of Space** 89

   Archytas at the Edge 89

   Is There Space beyond the Universe? 92

   The Illusion of Infinity 95

7. **Infinity and Paradox** 101

   Zeno: How the Tortoise Beat Achilles 101

   Two Responses to Zeno: Infinitesimals and Finitism 104

   Thomson's Lamp 107

   A Puzzle about Transition 111

   Democritus' Cone 115

   Atoms of Space and Time 119

8. Does Time Pass?   122

   The Mystery of Passage   122
   McTaggart's Proof of the Unreality of Time   127
   First Response: Presentism   135
   Second Response: the B-Theory   140
   Why Is There Only One Present?   143

9. The Cinematic Universe   148

   Muybridge's Horse and Zeno's Arrow   148
   No Motion at an Instant?   150
   No Motion in the Present?   156
   Zeno and the Presentist   159

10. Interfering with History   164

   The Lost Days   164
   The Alterability of the Past   167
   Dilemmas of the Time-Traveller   174
   Causation in Reverse   181

11. Other Times and Spaces   185

   Probability and the Multiverse   185
   Branching Space   191
   Objections and Consequences   193

12. The Arrows of Time   202

   The Hidden Signpost   202
   Three Arrows, and Why Things Fall Apart   206
   The Mind's Past   213
   The Seeds of Time   218
   Parallel Causes   221
   Is Time Order Merely Local?   223
   Are Causes Simultaneous with Their Effects?   226
   A Sense of Direction in a Directionless World   229

Concluding Thoughts   234

Mr Dunne's Dream and Other Problems   248

Further Reading   255

Bibliography   263

Index   269

CHAPTER I

The Measure of All Things

Philalethes: Our measurement of time would be more accurate if we could keep a past day for comparison with days to come, as we keep measures of spaces.

G. W. Leibniz, *New Essays on Human Understanding*

Incident at Greenwich

On the evening of 15 February 1894, a man was discovered in the park near the Royal Observatory at Greenwich in a most distressing condition: it appeared that he had been carrying or otherwise handling some explosive which had gone off in his hands. He later died from his injuries. The fact that he had been in Greenwich Park naturally provoked speculation: was he attempting to blow up the Observatory? Around this puzzling and ambiguous incident Joseph Conrad constructed, in *The Secret Agent*, a story of a double agent who had been instructed by a foreign power to blow up 'the first meridian'—i.e. the Greenwich Observatory—and so provoke outrage at what would be perceived as an attack on science or technology itself, the idea being that this would be a much more subtly unsettling attack on society than any assault on a prominent individual or group of innocent people.

By 1894, Greenwich had acquired a peculiar significance: it not only marked 0° longitude, it also stood for the standardization of time. For much of the nineteenth century, different towns in Britain kept their own time, and travellers from one place to another would often have to reset their portable timepieces on alighting. But the development of the railways made it increasingly important to dispose of these local variations, and 1852 saw the introduction of a standard 'Railway Time', as it was called. Finally, in 1880, Parliament passed the Definition of Time Act, which introduced a universal time, this being defined by the time on the Observatory clock at Greenwich. This, as we might imagine, could well have induced in some quarters the same resentment as the idea of a single European currency does in others today, though whether feeling ran sufficiently high as to motivate the blowing up of the Observatory is a matter for debate.

The idea of a standard time implies a standard timepiece, which raises the question of what it is for a timepiece to be entirely accurate. I discover that the grandfather clock is slow by noticing a discrepancy between it and my 1950s wristwatch. But on comparing my wristwatch with your digital watch, bought only last week, I discover that my watch is losing a few minutes every day. And were we to judge your timepiece against the standard of a caesium clock we should no doubt discover some further discrepancy. But this process must have a limit. Eventually, we arrive at a means of measuring time that we take to be as accurate as anything can be, and we take this to be our standard, according to which all other timepieces are to be judged. Now, does it make sense to inquire, of this standard, whether it is truly accurate? This may strike one as a strange question. Surely, one can ask of any means of time-measurement whether it is truly accurate or not, a truly accurate clock being one that judges two adjacent periods (for example, successive swings of a pendulum) to be of the same duration when and only when they are indeed of the same duration. But here we come up against a problem. There is simply no way of telling, for certain, that anything meets this requirement. We can only compare one timepiece with another.

How in fact do we decide that a particular means of measuring time is accurate (at least to an acceptable degree)? Here is one method. We can take a particular timepiece and reproduce it a number of times. We should take care to make the replicas as close to the original as possible in terms of their dimensions and physical composition. We should also ensure that they are placed in the same environmental conditions (one timepiece, for example, should not be in a much hotter place than another, or subjected to greater vibration, pressure, etc.). Then we should synchronize them and set them going. Finally, we should note whether they tend to stay synchronized, or whether they end up out of step with each other. If after, say, a few days they are very significantly out of step with each other, then we know that this particular means of measuring time is not particularly reliable. But if instead they remain perfectly in step with each other, if—to use the technical expression—they remain *congruent*, then we can be confident that we have found a reasonably accurate timepiece. The longer they remain congruent, the more accurate the method.

However, we can still, it seems, entertain the possibility of a timepiece remaining perfectly congruent with its replicas for years on end, and yet not keeping perfect time. For what it is to be completely accurate is not, we suppose, merely to be in agreement with a standard, but to measure time itself correctly. We take it for granted that the intention of the clock is to measure time. But perhaps this familiar notion is, after all, a rather peculiar one when we pause to scrutinize it. What is it for an instrument to measure *time*? The oddity of the idea of measuring time is well captured by a story of

O. K. Bouwsma's, 'The Mystery of Time (or, The Man Who Did Not Know What Time Is)'. The hero of the story is puzzled by clocks. He has been told that they measure time, but although he has seen them at work doing their measuring, he has not yet been able to see what it is that they measure. With other kinds of measuring instruments there is no problem. A tape-measure, for example, can measure a length of cloth, a pair of scales can measure a quantity of flour, a jug a volume of water, and so on. What is being measured in these cases is plain to see. But clocks seem to be able to register something that does not affect our senses at all. Perhaps there is some invisible, ethereal fluid flowing through these instruments, making the hands go around the dial? Or perhaps there is nothing at all, and the mechanism operates without any external prompting. Our hero begins to suspect it is all a con trick. In effect, the story is an inverted version of the Emperor with no clothes—there is in fact no trick: the clocks genuinely are measuring something that cannot be seen.

Initially, we smile at the man's innocence. He is simply mistaken in thinking that whatever is being measured must be capable of being seen. He takes the materialist outlook too far. But the problem goes rather deeper. The clock registers the passage of time in virtue of being a temporal process. We, too, register time's passing as we age. The cycle of the seasons is another change which cannot but indicate the rushing onward of that ever-rolling stream. Clocks are simply a peculiarly regular kind of change, and, like us, they register change by themselves changing. But now we begin to see what is so odd about the idea of clocks measuring time: unlike the kitchen scales, they are not entirely independent of what they are measuring. For when people talk of time, are they not simply talking, in an abstract way, of change? Consider how we experience time. I glance out of the window and see the branches of the horse chestnut gently swaying in the breeze; a bird alights on one of the branches for a moment and then flies off; I catch the droning of a passing motor car in the lane; and in the distance the striking of the church bell tells me that it is three o'clock. And if I shut my eyes and block my ears, I still register my fleeting thoughts. All these things press on me the passage of time. I experience time, in other words, through the experience of change. So perhaps time is neither more nor less than change. Moving objects, changing feelings, chiming clocks: all these *are* time. Or so we may naturally think. Hence the oddity of the idea that clocks measure time: they are what they measure. But then we return, and this time with greater puzzlement, to the question of what it is to measure time correctly—or *in*correctly. If a timepiece is measuring its own change, among other things, how can it get it wrong, so to speak?

At this point, we need to impose a little discipline on our rather rambling train of thought. Three questions need to be addressed: What is it for a timepiece to be accurate? Is time the same thing as change? What, if any, is the connection between these two questions? The remainder of this chapter will be concerned with the first of these questions. We will leave the others until Chapter 2.

Metric, Convention, and Fact

At one point in the preceding discussion we noted that, although we could perform a test that would show some kinds of timepiece to be more accurate than others, it was impossible to tell whether an instrument was 100 per cent accurate since all one had to judge accuracy by was other instruments, whose accuracy could always be called into question. We could not survey time itself, independently of all the changes that took place in time, to see if an instrument was measuring it correctly or not. Indeed, it was not entirely clear that we could make sense of time existing independently of these changes.

There are two responses to this limitation of ours. The first response is to deny that it is any kind of limitation at all. Once we have selected our standard—whether it be a caesium clock or a sundial—it makes no sense to ask *of the standard* whether it is accurate or not. It is accurate by fiat. The very act of selecting the standard confers upon it total accuracy. All instruments are judged by the standard: an instrument is accurate if and only if it agrees with the standard. It follows from this that the standard cannot fail to be accurate (or perhaps that it makes dubious sense to apply the concept of accuracy to the standard). Since the choice of standard is a matter of convention, though not completely arbitrary convention, this viewpoint is known as *conventionalism concerning the metric of time*. It is important to distinguish this rather controversial position from the obvious and unexciting observation that our choice of a particular unit with which to measure time is a conventional matter. The fact that we divide up days into hours, minutes, and seconds, for example, is not something divinely ordained, but simply a matter of convenience. (It is interesting to note that in this age of decimalization, where pounds and ounces have given way to grammes, yards to metres, and—in Britain—shillings to five pence pieces, we still have twenty-four hours in a day. There have, however, been attempts to impose a decimal system on time—in France after the Revolution, for example—and some decimal clocks still exist, although confined to museums.) What, according to conventionalism, is a matter of convention is not merely the choice of unit, but whether two given successive intervals of time are of the same duration or not. Take your last three heartbeats. Was the interval between the second and third the same as that between the first and second? According to one method of timing those intervals, they were the same; according to another, they were not. But were they *really* the same? The conventionalist has no use for such a question. It is like asking whether this toadstool is *really* poisonous: it will be so for some creatures, but not for others. (Of course, whether a toadstool is poisonous for a given creature is not merely a matter of convention!)

So much for the first response to our inability to compare a timepiece directly with time itself to check its accuracy. What of the second response? This is *objectivism concerning temporal metric*, and for it the question about heartbeats does have a point. Whether or not we can discover if two successive intervals were equivalent, there is, for the objectivist, a fact of the matter as to whether they were equivalent, independently of any conventional choice of a measuring system. (By 'measuring system' is meant here a means of comparing the intervals, not a unit of time.) As objectivism is sometimes expressed, time has an *intrinsic metric* or measure. Again, this does not mean that time is *really* divided up into hours and we have happily stumbled upon the right unit of measure, but rather that, independently of any way of determining it, *this* pair of successive intervals is a pair of equivalent intervals, and *that* pair is not. One of the consequences of objectivism is that there are some facts in the world that remain 'secret'. That is, we can never know for sure whether or not they really exist. And this is not just an accidental limitation, something we could perhaps overcome if we put our minds to it, but something that could not possibly be otherwise. For some people, this is an unacceptable consequence. There is a theory that at one stage commanded a wide following in philosophy, and in one form or other still has its adherents, and that we may characterize as follows: if a given statement about the world is such that we have no possible way of discovering whether or not it is true, however ideally we are situated, then that statement is neither true nor false. This position is one version of *verificationism*. Some verificationists wanted to go further and say something rather stronger: that the statement in question is actually *meaningless* in those circumstances. For the time being, however, we can confine ourselves to the more modest position. To express that position slightly differently: there are no facts about the world which we could not, at least in principle, discover to be the case. No fact is essentially impossible to discover. To the extent that objectivism is committed to such undiscoverable facts, objectivism would be rejected by anyone subscribing to a verificationist view of truth.

Conventionalism offers some comfort to Bouwsma's hero. It tells him that he need no longer search for the elusive thing that clocks measure, for in reality they are measuring nothing. That is, there is no objective feature of the world that is being monitored by the clocks. They are simply mechanisms by which we may order our lives, which enable us to synchronize our meetings, and which may remain or fail to be congruent with each other. That is the end of the matter. I suspect, however, that our intuitions point us in the direction of objectivism. In which case, should we not be worried by the inaccessible facts which objectivism says are out there? Is there no room for the suspicion that, since we have no use for such facts as the putative fact that these two intervals are really equivalent (though we can never prove it) we may as well do away with them?

Those are not meant to be rhetorical questions, whose answer is obvious; they simply raise a worry over what otherwise seems to be the natural assumption that the metric of time—how long things take—is a matter of fact rather than convention. How might we address this worry?

Time and the Laws of Nature

Making a number of exact reproductions of a timepiece and seeing if they remain in step is one way of testing the accuracy of a particular means of time measurement. This method, as we have seen, is ultimately inconclusive. There is, however, another, and that is to see whether or not the deliverances of our timepiece are consistent with the laws of motion. Consider the following:

> A body continues in a state of rest or uniform motion unless acted upon by a force.

> The acceleration of a body is a function of its mass and the force acting upon it, such that Force = Mass × Acceleration.

These are, respectively, Newton's First and Second Laws of Motion. Both of them make implicit reference to the notion of equal intervals of time. For a body is moving with constant velocity if it covers the same distance in each of a series of equivalent intervals (e.g. a foot every second), and accelerating (or decelerating) if it covers a greater (or smaller) distance in each successive and equivalent interval. We can then set up an experiment in which the rate of motion of a body is measured by the timepiece whose accuracy we wish to test.

Suppose, then, we attach, to a body of known mass, a spring. By means of this spring we can drag the object along the ground at varying speeds. The force exerted on the object can be measured by the extent to which the spring is extended. Having calibrated the spring appropriately, and marked out a series of equivalent distances along which the object is to be propelled, we select the timepiece to be tested. We then conduct a series of tests, exerting a variety of forces on the object, causing it to move at various speeds, and at various rates of acceleration, measuring motion all the while by means of the selected timepiece. We end up with a long series of pairs of values for force and acceleration. If the results are consistent with the previously accepted Laws of Motion, then our timepiece may be considered at least approximately accurate. If the results are not consistent, then we may regard it as inaccurate.

Of course, the Laws will in the first place have been established through the use of timepieces, so our faith in them is not independent of means of measuring time. Nevertheless, the point is that there is a connection between the way in which we measure time, and the laws we discover to govern motion. Some timepieces will point the way to relatively simple laws relating motion and force, some to much more complex ones, and others to no systematic relationship at all. We naturally assume that the world is an ordered place, and that there will be systematic relationships between the forces acting on a body and the subsequent motion of that body. And we will naturally prefer simpler relationships to more complex ones. So we have a way of choosing between competing timepieces. (Of course, the crude set-up described above will not be adequate for distinguishing between timepieces which differ from each other only slightly, but then other tests could be devised.)

This connection between our measurement of time intervals on the one hand and the laws of motion on the other was pointed out by the great mathematician Leonhard Euler (1707–83) who held a chair first of physics and then of mathematics at St Petersburg's Academy of Sciences. Euler is regarded as the founder of a branch of mathematics known as *topology*. His suggestion, in his *Reflections on Space and Time,* was that a given cyclical process can appropriately be described as *periodic* (i.e. each cycle in the process takes exactly the same time as every other) if, having defined a unit of time in terms of that process, we find that Newton's First Law of Motion is confirmed.

This may offer the objectivist over temporal metric some ammunition against the conventionalist. First, the fact that we can test the accuracy of different timepieces against what we take to be the laws governing motion suggests that, after all, facts concerning temporal metric may not be inaccessible. So conventionalists are not necessarily on firm ground when they say that objectivists are obliged to suppose that there are facts we can never discern. Secondly, taking the laws of motion to be truly descriptive of the world actually entails objectivism over metric. If it is an objective fact that an object of such-and-such a mass will be accelerated by exactly this much when such- and-such a force is applied to it, then it is an objective fact that some successive intervals are longer or shorter than, or the same as, others.

These are certainly significant considerations, but they are not decisive. To take the first point, our experiment only succeeds in testing the accuracy of a timepiece if we can be sure that our measurements of force and distance are accurate. But we are not entitled simply to assume this. How did we measure equivalent distances? Perhaps by means of a ruler. This certainly makes the measurement of space easier than the measurement of time, since we cannot transport a temporal 'ruler' (such as the interval between two strikes of a bell) from one time to another. But it is no more secure. Did the ruler remain exactly the same size as we moved it from one place to another? We cannot be entirely sure. Given this uncertainty, what we are testing in the experiment is not simply the accuracy of a timepiece, but the accuracy of our whole measurement system. And even if we get the result we wanted, we cannot be sure that the individual measurements are accurate, because it may be that some deficiency in our time measurements are compensated for by deficiencies in our distance or force measurements. However, this possibility becomes more remote the more tests we perform.

The second point against the conventionalist—that taking the laws of motion as objectively true entails objectivism over metric—is more significant, but the conventionalist can simply reply that the laws of motion themselves are simply conventions. They are undoubtedly extremely useful, in that they correctly predict the observed outcome of our experiments with motion, but their usefulness does not entail their truth. When playing chess against a computer, we find it useful to ascribe states of mind to the computer: it *intends* to take that bishop, it is *willing* to sacrifice that knight, it *knows* I wish to attack its queen, etc., etc. Is the computer really in any of these states of mind? Has it a mind at all? Most people would be inclined to answer 'no' to this question, but still find it useful to treat the computer *as if* it had a mind in predicting its moves, this being a much easier method than trying to work out what it will do from a study of electrical impulses through its circuitry. This, the conventionalist may insist, is how things are with the laws of motion. It is convenient to take them as describing things out there in the world around us—equal intervals, equal distances, equal forces—but they are not actually doing so. How plausible is this? The notion of quantity is a very basic, indeed indispensable one, without which we cannot adequately describe our world, act in it, or predict what will happen. It is very hard indeed to believe that a quantitative conception of things does not pick out aspects of reality itself, aspects that would be there whether or not anybody was describing them.

Conventionalism about temporal metric is a view specifically about time, of course; it does not say anything explicitly about other quantities. But what we have seen is that it is not possible to think of conventionalism as applying only to time: it has further, and far-reaching, consequences. We have to treat the laws of motion—*any* laws of motion, not just Newton's—as not truly descriptive of the world, not because they are merely approximations, but because they simply fail to pick out genuine features of the world. Conventionalism, then, is a very bold position indeed.

Someone will complain that so far we have simply been skirting around the question that so puzzled Bouwsma's hero: what *is* it that clocks measure? It is time to tackle this question head on.

Questions

Why might people have found the idea of a 'universal time' so objectionable? Does this tell one anything about how they conceived time?

What would you say to Bouwsma's man who did not know what time was?

If there is no objective truth of the matter as to whether one event is longer than a subsequent event, how is it that some processes are better at measuring time than others?

CHAPTER 2

Change

> Stand still, you ever-moving spheres of heaven.
> That time may cease, and midnight never come
>
> Christopher Marlowe, *Doctor Faustus*

Time as Change

'Stop all the clocks,' writes W. H. Auden, in a poem lamenting the death of a (real? imaginary?) friend. Grief brings time to a standstill, or rather brings with it a desire to stop time taking us further and further away from the person we loved. A different kind of grief, and one that blights her embittered existence, is suffered by Miss Havisham, the elderly lady whom Pip visits in Dickens's *Great Expectations*. Abandoned by the groom on the day they were to be married, she sits, a grotesque figure in her wedding-dress, in a darkened room, with the clocks stopped forever at twenty minutes to nine. At first, Pip feels only horror, until he begins to take in the significance of his surroundings:

It was then I began to understand that everything in the room had stopped, like the watch and the clock, exactly, a long time ago. I noticed that Miss

Havisham put down the jewel exactly on the spot from which she had taken it up. As Estella dealt the cards, I glanced at the dressing-table again, and saw that the shoe upon it, once white, now yellow, had never been worn. I glanced down at the foot from which the shoe was absent, and saw that the silk stocking on it, once white, now yellow, had been trodden ragged. Without this arrest of everything, this standing still of all the pale decayed objects, not even the withered bridal dress on the collapsed form could have looked so like grave-clothes, or the long veil so like a shroud.

As all this decay shows, attempts to stop time are futile, for we cannot stop change. But what if all change *were* to stop? Would that be the end of time too?

The question is raised in Aristotle's *Physics*, which contains one of the earliest and fullest attempts to provide a philosophical account of time. Aristotle (384–322 BC) is a towering figure in the history both of philosophy and science. His writings cover an extraordinary range of subjects, from biology to the nature of tragedy. But we should not be misled by the title of the *Physics* into thinking that Aristotle is engaged in the kind of study that we would recognize as physics in the modern sense. He is not, for example, concerned to formulate laws governing physical phenomena. Rather, he is interested in the most general categories into which physical items, properties, and phenomena fit—change, objects undergoing change, places, time, quantities—and attempts to give an account of what these are, and how they are related to each other. The kind of inquiry to be found in the *Physics*, then, is a highly abstract one, but one that Aristotle thinks necessarily precedes any more specific investigation into the detailed workings of physical phenomena.

Aristotle's discussion of time begins with a catalogue of paradoxes that seem to point to the conclusion that there is in reality no such thing as time. It is remarkable how many of Aristotle's successors embraced that extraordinary conclusion. Aristotle himself, however, is not willing to deny the reality of time, and hopes later on to resolve the paradoxes (and we shall encounter some of them ourselves later in this book). Having alerted us to the dangers of the territory through which we are now journeying, Aristotle presents the dominant view of his predecessors, which was that time and change are one and the same. The most famous exponent of this view was Aristotle's one-time teacher Plato (*c.* 429–347 BC, who presents us, in the *Timaeus* (which takes the form of a dialogue between Socrates and three friends), with an account of the 'birth of time', this being the first motion of the heavenly bodies. But, as with many of his predecessors' opinions, Aristotle finds fault with this identification of time and change. Time could not be the same thing as change, he says, for first change can go at different rates, speed up or slow down, but not so time, and secondly change is confined to a part of space whereas time is universal.

What are we to make of these objections? Surely time *does* speed up or slow down, or at least it appears to do so. For people in love, a few hours spent together will pass all too swiftly, whereas time will hang heavily during a labour of unremitting tedium. But such phenomena are easily dismissed as illusory. We can be deceived about spatial matters, such as the shape or size of an object, or its distance from us, so why not also about temporal matters? To see whether it makes sense to suppose that *time itself* could pass at different rates, consider how we measure the rate of other kinds of change: the speed of a passing bus, for example. We measure the distance the bus covers against time. Or consider a kettle on a stove. Its rate of heating is given by measuring the rise in temperature against time. So rate of change is variation in some dimension in so many units of *time*. How, then, would we measure the rate of passage of time? Why, against time, presumably. But this leads to the conclusion that the rate of the passage of time must never vary. For how long could five minutes take if not five minutes? But Aristotle's objection perhaps misses the point. It is true that time could not be identified with *particular* changes, such as the crumbling of a sand castle. But to identify time with change is surely to identify time with change *in general*. Now it is not at all clear that change in general—that is, the sum of all changes in the universe—could intelligibly be regarded as proceeding at varying rates. Try to imagine every change in the world suddenly doubling in speed. Does that idea make sense? Aristotle would not have thought so. For one thing, we could not possibly *notice* such a change in rate, for we only notice the change in the rate of some change when comparing it with other changes. We notice the shortening of days with the onset of winter by measuring the time between sunrise and sunset against conventional timepieces or our own biological clocks.

The idea that time is to be identified, not with particular changes, but with change in general seems also to avoid Aristotle's second objection, that change is confined to parts of space, whereas time is universal. Only individual changes are spatially confined, but the totality of change covers the whole of space.

This may have removed one ambiguity in the notion of change, but there remains another. What kind of change do we suppose time to be? Do we think that time is the same as the sum of all the ordinary changes of which we can directly be aware, such as the changing colour of a leaf, and also those which underlie perceivable changes, though not themselves perceivable, such as the motion of molecules? Or are we instead thinking of the passage of time itself, the inexorable movement of things once present into the ever-distant past? Of course, a philosopher who said that time was to be defined as the passage of time would not get much of a following, since such a definition defines time in terms of itself. We need to have some way of defining the passage of time. This is most naturally described (though some philosophers would object to this way of describing it) as the change in events as they cease to be future, become present, and then increasingly past.

One way of capturing the distinction above is in terms of first- and second-order change. *First-order change* is change in the properties of things in the world, where 'things' are conceived of as items that persist through time, such as trees, atoms, and persons. First-order changes, then, are what we would ordinarily describe as events. Can events themselves change? *Second-order change* is, or would be, the change that events suffer ('second-order' because it is a change in first-order changes) as they cease to be present and slip into the ever more distant past. Second-order change, then, is nothing other than the passage of time itself. So when we say time is change, is this first-, or second-order change? Or both? It is certainly a plausible suggestion that time is at least partly constituted by second-order change, for how could time exist unless it also passed? The passage of time is surely its most striking characteristic. Nevertheless, in a later chapter we shall consider reasons to favour the strange hypothesis that, although time is real, it does not pass but exists, as it were, in a 'frozen' state.

In the remainder of this chapter, however, we shall concern ourselves with the suggestion that time is *first*-order change. Now, it might seem to us that we could imagine every process in the universe coming to a stop—perhaps after the so-called 'heat-death' of the universe, where all energy is perfectly evenly dispersed—and yet time continuing to pass. Endless aeons of time might pass in a completely dead, motionless universe. So maybe time can exist in the absence of first-order change. But is this a real possibility?

Time without Change?

Aristotle did not think time without (first-order) change was a real possibility, and many later writers agreed. We shall now look at three arguments against the possibility of what we might call a 'temporal vacuum', that is a period of time in which absolutely nothing happens. Aristotle's is the simplest: were all change to cease, we would cease to notice the passage of time. This statement is, of course, irrefutable. To notice anything is to undergo a change in mental state. The cessation of all change is also the cessation of any experience, so it is impossible to experience a temporal vacuum (in the sense of experiencing anything *as* a temporal vacuum). But for this irrefutable proposition to imply that it is impossible for time to continue in the absence of change, some connection must be made between the reality of something and the possibility of experiencing it—or at least having evidence that reality contains the feature in question. The missing link is provided by the stronger verificationist principle introduced in the previous chapter: if there is, even in principle, no possible means of establishing whether a statement is true, or at least likely to be true, then that statement has no meaning. The resulting argument we may call 'the experience argument'. In full, it goes as follows:

The experience argument

1. During a period of time without change, there would be no experience at all—since experience itself is a form of change—and so no experience of the period *of* time without change.

2. A period of time by itself changes nothing, and so makes no difference to what we could experience after that period.

3. We can establish that some contingently true statement is true only if its being true could make some difference to what we experience, either now or at some later stage.

Therefore:

4. We cannot possibly establish that a period of time without change has occurred.

5. If it is impossible to establish whether or not some statement is true (or likely to be true), then that statement has no meaning.

Therefore:

6. Any statement to the effect that a period of time without change has occurred would have no meaning.

A 'contingently true statement', mentioned in premiss 3, is one that concerns a state of affairs that might not have obtained, such as 'I like asparagus.' It is to be contrasted with a necessarily true statement such as 'If I like asparagus then I like asparagus', which could hardly have failed to be true.

Once the argument is stated as explicitly as this, it begins to seem less compelling (although it should be said that this is not the only way of enlarging upon Aristotle's own rather brief rejection of the temporal vacuum). For one thing, the conclusion just seems far too strong. Can it really be *meaningless*—as opposed to just wrong—to say that a temporal vacuum lasting five minutes has just occurred? We certainly seem to know what it means, namely that for five minutes absolutely nothing happened in the entire cosmos. And if the statement really is meaningless, despite appearances to the contrary, then it is hard to see how we are supposed to understand premiss 1 of the argument, which tells us how things would be *if* there were a temporal vacuum. How can it be both meaningful and true to say that there would be no experience in and of a temporal vacuum if all talk of a temporal vacuum is meaningless? The conclusion of the argument, then, appears to undermine its first premiss. That can hardly be a mark of a good argument.

A point in favour of the meaningfulness of temporal vacuum talk is that, although we cannot tell whether the statement 'There was a temporal vacuum going on' is true or false, we can recognize that the statement 'There *is* a temporal vacuum going on' *is false.* Could we do that if the statement had no meaning? Surely genuinely meaningless statements are neither true nor false, since they fail to say anything intelligible about the world.

The suspicion, then, is that the principle of meaningfulness articulated in premiss 5 is far too strong. There are, however, occasions when it delivers the right result. Suppose I say to you 'There is a moncupator in my fridge', and when you ask, reasonably enough, what a moncupator is, I confess that I can offer very little in the way of characterization. For moncupators are invisible, do not prevent other objects from occupying the same region of space, and in general are completely undetectable. There is, in other words, no possible means of establishing whether or not there is a moncupator in my fridge. At this point, you would be justified in declaring the word 'moncupator' to have no meaning, for there is no way in which we can learn to use it correctly. We cannot, for example, teach a child the meaning of the word by pointing and saying 'Look, a moncupator!'

In consequence, my statement about the contents of my fridge is similarly devoid of meaning. But we are not in anything like the same sorry state when it comes to the word 'time'. *This* word is one that we successfully learn to use appropriately (although it is no simple task to explain in detail *how* we come to use it appropriately). So perhaps it does not matter if we cannot establish whether the statement 'There was a period of time in which nothing happened a while ago' is true or false: it is meaningful because the component parts are meaningful. But this is too generous a criterion of meaningfulness. Consider 'Happy the is a whereas pink'. The individual words have a meaning, but the whole does not. We could insist that the statement be grammatically well-formed, but this is still not enough. Consider the grammatically well-behaved but ultimately unintelligible 'No probable tree vetoes a pungent absence.' What is wrong with this? Well, it is pretty clear that we have no idea of how to go about establishing whether it is true or false, but if this is to be our criterion then we are back with what we suggested a moment ago was an implausibly strict condition of meaningfulness.

Perhaps a more promising approach is this. When we learn a word, we learn the contexts in which it may appropriately be applied. We do not, for example, think it appropriate to apply colours to numbers ('Three is a red number'). So one condition of the meaningfulness of a sentence is that it does not use terms in inappropriate contexts. Here is an interesting and relevant example from the Austrian philosopher Ludwig Wittgenstein (1889–1951), who spent most of his professional life in Cambridge: 'It is 5 o'clock on the sun.' Now is the sun a place where it would be appropriate to talk of its being 5 o'clock? Arguably not, for what time it is depends upon one's time zone, which in turn is defined by one's longitude— a position which is well-defined for the surface of the Earth but not elsewhere. However one could turn 'It is 5 o'clock on the sun' into a meaningful statement by means of further explanation, for example that sun time is defined as Greenwich Mean Time. The point is that 'It is 5 o'clock on the sun' only means something when it is clear how the sun can be included among the legitimate contexts in which dates and times are applicable.

Now, when we learn the concept of time, we necessarily do so in a context of change. We learn, for example, that it is appropriate to say that time has passed when we have witnessed some change. The different times of day are often introduced by association with different activities (1 o'clock: lunch; 6 o'clock: bath-time, etc.), and the different months with appropriate seasonal weather. Further, when we think, not of a repeatable time like 6 o'clock, or autumn, but a particular, unique and unrepeatable moment, we inevitably think of it in terms of what was happening at that moment. And, as we noted in the previous chapter, time is measured in terms of periodic changes. So when we contemplate the phrase 'period of time in which no change occurs', we find a term (time) which is being used in an unfamiliar context. We might go further, and say that the term is actually inapplicable in such a context: where there is no change, we cannot talk of time passing. Talk of a period of time in which no change occurs therefore makes as little sense, so the argument goes, as talk of a number changing colour or of its being 5 o'clock on the sun.

This is a rather more subtle argument than the one that goes, in effect, 'we could never notice time passing if there were no change, so time does not pass in the absence of change'. That way of putting it makes the argument from experience look rather obviously fallacious. However, even the more subtle argument can be resisted. It is clearly true that we acquire temporal concepts in a context of change. Indeed, this is true of any concept, since we are constantly surrounded by change. What is not at all obvious is that temporal concepts are therefore inextricably linked with the idea of change. We can, after all, abstract the idea of a time from the particular events that occurred at that time. I may, for example, be contemplating a particularly poignant moment in the past when I was saying goodbye for the last time to a friend at a railway station. I recall to mind lowering the window of the carriage, seeing the station clock indicate the time of departure, hearing the stationmaster slamming the train doors, catching the acrid whiff of smoke from the engine (the memory is evidently a rather old one), and whispering 'Try to forget me'. As I remember the scene, I think 'If only I had at that moment leapt off the train, flung myself on the platform and cried "Marry me!"' Had I done such a thing, that moment, indeed perhaps all of life thereafter, would have been different. So, if I can imagine a time without being obliged to think of it as being filled by the very events that did happen then, why can I not think of it as filled by *no events at all*? Why can I not say 'If only the entire cosmos had come to a standstill for five minutes at that moment'? The experience argument, then, whether in its original or more subtle form, rests on questionable assumptions.

Let us now turn to a second argument against temporal vacua, one put forward by the German philosopher, G. W. Leibniz (1646–1716). In the *New Essays on Human Understanding*, which is written in the form of dialogues, Leibniz has one of his characters, Theophilus, say the following:

[I]f there were a vacuum in time, i.e. a duration without change, it would be impossible to establish its length. . . . we could not refute anyone who said that two successive worlds are contiguous in time so that one necessarily begins as soon as the other ceases, with no possible interval between them. We could not refute him, I say, because the interval is indeterminable. (Remnant and Bennett 1981, 155)

This passage suggests an argument against the temporal vacuum, although the conclusion is not actually made explicit. The interpretation is a plausible one, as we know from things he says elsewhere that Leibniz did indeed reject temporal vacua. Laying bare the tacit assumptions, then, we have the following:

The measure argument
  1. Periods of time are measured by changes.
  Therefore:
  2. Since, by definition, nothing happens in a temporal vacuum, there is no possible means of determining its length.

3. If there is no means of determining the length of a temporal interval, it has no specific length.
4. Every interval of time has a specific length.
Therefore:
5. There cannot be a temporal vacuum.

Premisses 1 and 2 look safe enough, but 3 seems to contain a hidden fallacy. As with Aristotle's argument, a connection needs to be made between our inability to discover a certain kind of fact and the proposition that there are no facts of that kind to discover. In this, as in the earlier, case the connection must be some theory of meaning. The principle that meaningfulness depends on verifiability would provide the desired result, but as we noted above, the principle seems implausibly strong. As before, however, we can appeal to more subtle considerations: given judgements concerning the length of intervals are made in the context of regular changes, we cannot legitimately extend the notion of duration to the context of no change. If this is the right way to defend the measure argument, then it is clearly not really distinct from the experience argument, but is just a slightly more specific version of it. It therefore does not seem to merit a separate response. There is, however, one point which applies specifically to the measure argument, and which is sufficient to defeat it.

The problem concerns the tension between premisses 3 and 4. Our view of these will depend on whether we adopt the conventionalist or objectivist view of temporal metric, the debate between which was described in the previous chapter. Take a look first at premiss 3: 'If there is no means of determining the length of a temporal interval, it has no specific length.' This is strongly suggestive of conventionalism, which states, we recall, that whether or not two successive intervals are the same or not depends on one's chosen means of measuring them. In the absence of any possible means of measurement, there will simply be no fact of the matter as to whether the two intervals are the same or not, and so no fact of the matter as to how long each is. The conventionalist, then, will happily accept premiss 3. The objectivist, however, who holds that the length of an interval is an objective matter, and not dependent on the availability of any means of measurement, will certainly reject it. Now consider premiss 4: 'Every interval has a specific length.' The objectivist will have no difficulty with this—indeed, it might be taken as a statement of objectivism in a nutshell. For this reason, however, the conventionalist will reject it. For conventionalism allows for the possibility of intervals which do *not* have specific lengths, precisely in those cases where there is no possible means of measuring them. So, in sum, whether we are conventionalists or objectivists over temporal metric (or still sitting on the fence) we have good reason not to accept *both* premiss 3 and premiss 4, although we might accept either one.

And so we may take our leave of the measure argument.

There is something, as we might put it, rather egocentric about the two arguments we have just considered: both are concerned with what we can experience or be affected by. Let us now remove ourselves from the picture, and ask what effects temporal vacua could have on the world at large.

Everything Has a Reason

This takes us to our third argument, which is suggested by a characteristically terse and enigmatic remark from the Ancient Greek philosopher Parmenides:

And what need would have aroused it later or sooner, starting from nothing to come into being? (Barnes 1982, 178)

The 'it' here is the cosmos, or entire universe. It cannot have simply come into being at a particular moment in time, suggests Parmenides, for not only would there would have been nothing to make it come into being, we would have no explanation of why it came into being at precisely that moment and not at an earlier or later time. The principle on which this line of reasoning rests is rather more visible in the following passage from Leibniz's reply to a letter written by the Revd Samuel Clarke:

> For since God does nothing without reason, and no reason can be given why he did not create the world sooner; it will follow, either that he has created nothing at all, or that he created the world before any assignable time, that is, that the world is eternal. But when once it has been shown, that the beginning, whenever it was, is always the same thing; the question, why it was not otherwise ordered, becomes needless and insignificant. (Alexander 1956, 38–9)

Here Leibniz is attacking those who take time to be independent of the existence of the various changes that constitute the history of the cosmos. For suppose that time were so independent, and that the universe had a beginning in time. Then there would have been aeons of empty time prior to creation, and no answer to the question: 'Why did God create the universe at precisely that moment and no other?' for one moment of empty time is much like any other. However, once we recognize that the beginning of the universe and the beginning of time are one and the same, God is no longer faced with such a choice. The universe could not have begun any earlier or later than it did, since there was no time before the universe existed. The principle appealed to, and which here is stated in theological terms, is called the Principle of Sufficient Reason. It has an appeal even for those who do not believe in a deity, but who do believe that everything has a cause. So putting the principle in more neutral terms, terms which are consistent with, but which do not imply, the existence of God, we have something like the following: for everything that occurs at a given moment, there is always an explanation of why it occurred at precisely *that* moment and not at some other moment. This not only makes difficulties for the idea of empty time before the existence of the universe; it also makes difficulties for the idea of limited periods of empty time during the history of the universe. For suppose that, exactly one hour ago, there was a complete temporal vacuum in which everything stopped for ten minutes. What made everything start up again? Since things remained exactly the same during that ten-minute period, there is no explanation of why things started up after ten minutes rather than, say, five, or fifteen.

Laying, once again, all assumptions bare, the third argument can be presented thus:

The sufficient reason argument

1. If there have been temporal vacua in the past, then there have been times when change has resumed after a period of no change.

2. For every change that occurs at a given moment, there is always an explanation, in terms of an immediately preceding change, of why it occurred at precisely that moment and not at some other moment.

Therefore:

3. There is no explanation of why a change occurring immediately after a temporal vacuum occurred when it did (since no change immediately preceded it).

Therefore:

4. There have been no temporal vacua in the past.

We may note at once that the argument does not rule out a temporal vacuum in the *future*, though it would have to be one that had no end: an eternity of darkness and silence after the death of the cosmos. Not, it has to be confessed, a cheering thought. This is a serious limitation of the sufficient reason argument, though we could supplement it with the following considerations: just as everything has a cause, everything has an effect. Change thus begets change, so once started on its changing course the universe could never cease changing. This further line of reasoning, however, is not hard to resist. Even if everything has an effect, why should the effect take the form of change, rather than the cessation of change? Could some changes not cancel each other out? Now, if a temporal vacuum—the absence of change—can be an effect, why can it not be a cause? We cannot say that it is cause of *change,* for otherwise we run up against the sufficient reason argument. But we could say that the absence of any change whatsoever in the cosmos is the cause of the *continued* absence of change. This suggests a compromise in the debate over vacua. Before revealing that compromise, however, let us sum up the situation so far, and introduce some useful terminology.

We began this chapter considering the suggestion that time and change are one and the same. This is one form of what is referred to as *relationism about time.* This brand of relationism asserts that time is just an ordered series of events, each individual moment identified with a collection of simultaneous events. Opposed to relationism is *absolutism about time,* which regards time itself as existing quite independently of what is happening *in* time. The absolutist will affirm the possibility of temporal vacua, for if time is independent of change it can exist where there is no change at all. The relationist, typically, will deny the existence of temporal vacua. Now, one way of supporting relationism is by means of arguments designed to establish the impossibility of temporal vacua, and we have considered three such arguments. Two of them, unfortunately, rest on dubious premises, or at least on dubious combinations of premisses. The third, although perhaps more persuasive, was rather more restricted in the scope of its conclusion. But perhaps we can develop it a little further.

One way of expressing the sufficient reason argument is to say that temporal vacua do not explain anything. They make no difference. And it is a powerful argument for something's non-existence that it would make absolutely no difference, as the example of the moncupator shows. It is a problem for the absolutist to explain what difference time makes *over and above* the effects of what happens in time. The egg is ready to be consumed five minutes after I started to boil it, not simply because five minutes have passed, but because of the changes going on inside the egg throughout that time. Again, it is not time that heals all wounds, but rather the (psychological, physical, political) changes that go on in time that do so. On the other hand, if we allow the absence of change to be a cause, then it is not true to say that temporal vacua could never make a difference. The fact that nothing whatsoever was happening at one time would explain why nothing whatsoever was happening at a later time. This is not, admittedly, the most exciting of explanations, but it does appear to satisfy the demand for sufficient reason.

The absence of anything happening is not the same as there being nothing at all: the absence of change is a state of affairs, and states of affairs, like events, occupy times. In fact, if one is going to be precise about this, one could insist that, fundamentally, all that exists at individual times are states of affairs, and what we call events are just series of different states of affairs (a view of change we shall consider critically in a later chapter). For example, the event that consists of a commuter travelling to work is just a series of states in which the commuter occupies a number of positions vis-à-vis other objects. So a temporal vacuum is a series of states of affairs, distinguishable only in that some states of affairs occur later than others. And this is what provides us with a face-saving manoeuvre for both relationist and absolutist, allowing them to retire from the field of combat with equal honour. The relationist can continue to maintain that time does not exist independently of its contents, but make the important concession that these contents do not have to be viewed as events only: they might be unchanging states of affairs. This then allows what the absolutist has insisted on all along, namely the possibility of time without change.

Questions

Why does it not make sense to think of *time* speeding up or slowing down?

Does it make sense to imagine every process in the universe suddenly doubling in speed?

Can you imagine any circumstances in which it would be possible to measure the length of a temporal vacuum?

CHAPTER 3

A Box with No Sides?

Companion: Good, now there'll be more room in the car.
H. W. B. Joseph: Well, no, surely: not more room, only less of
 it occupied!

Quoted by Geoffrey Madan, Notebooks

Where Two Worlds Meet

In May 1643, the French philosopher and one-time soldier René Descartes (1596–1650) began a long correspondence with Princess Elizabeth of Bohemia. In her first letter, of 6 May, Elizabeth asked him how, if the soul and the body were so different in their natures, the soul could influence the actions of the body. The problem was an acute one for Descartes, as he took extension—that is, having breadth, depth, and height—as being the sole defining characteristic of material things. The soul or mind he took to be unextended. Its defining characteristic was that of being a thinking thing. But how can something non-spatial interact with something spatial? His reply to Elizabeth of 21 May was somewhat evasive: he did not actually undertake to explain how the two interacted, but merely warned against taking the inter-action with which we are most familiar, that between material bodies, as the model for that between soul and body. Not surprisingly, Elizabeth continued to pursue him on the matter.

A few years later, Descartes was corresponding with Henry More (1614–1681), a young Fellow of Christ's College, Cambridge, and later one of a group of theologians known as the Cambridge Platonists. More objected to Descartes's characterization of the difference between soul and body. Mind, soul or spirit (and therefore also God) could, suggested More, occupy space as much as matter; what distinguished the latter was not extension but impenetrability: the fact that a piece of matter would prevent another piece of matter from occupying the same place. For More, space is a real thing that would exist in the absence of matter. It provides the medium where two worlds—the world of spirit and the world of matter—meet. Space also has a theological significance for More: its existence is explained by the fact that it is an attribute of God. More was apparently influenced by Jewish writings in which God appears to be identified with space (the word *makom is* used to denote both God and place). And so we have an explanation of the doctrine of Divine Omnipresence, which crops up numerous time in the Old Testament: 'Can any hide himself in secret places that I shall not see him? saith the LORD. Do I not fill heaven and earth?' (Jeremiah 23: 24).

If space is an attribute of God, then it is never truly empty, although it can be empty of any material thing. Yet, in More's time, many thinkers (including Descartes) were deeply opposed to the notion of the void, that is, space without body. To find the origins of this opposition, we have to go back to Ancient Greece.

Aristotle against the Void

If the idea of empty time is both unfamiliar and contentious, the idea of empty space is neither. We cannot look at the world without seeing the distances between objects. What are these distances but empty space? Not quite empty, someone will object, as they are filled with air. But air itself is full of empty spaces, for it is composed of molecules of nitrogen, oxygen, carbon dioxide, water, and the inert gases (among other things), and these are not packed together but are far apart and constantly moving. Most of the atmosphere, it seems, consists, literally, of nothing at all. As one of Aristotle's predecessors, Democritus, put it, the world is nothing but atoms in the void. These two notions—of the ultimate constituents of matter, and of empty space—go together, historically if not logically. For the atomists, from Democritus to Dalton, saw that gaps between atoms (or molecules) provide a natural explanation of the difference between the various phases of matter: gaseous, liquid, and solid. As a solid melts, the molecules move both further apart and more rapidly. As the liquid evaporates, the molecules move yet further apart and more rapidly still. That is why a certain volume of water will take up a much greater volume when it is converted into steam. It is also why substances tend to mix more readily as gases than as liquids.

But Aristotle rejected both atoms and void, and physics from his day to the time of Galileo was essentially an Aristotelian physics. Nature abhors a vacuum, as Aristotle's famous dictum had it. For Aristotle, all matter is continuous: it can be divided up indefinitely (in principle, at least) and there are no gaps in it. Moreover the cosmos itself, though finite, has no empty space around it. The difference between gases and liquids is simply explained by the 'finer texture' of the former (that this texture is something that itself stands in needs of explanation seems not have worried Aristotle). So all space is completely filled. The idea of empty space troubled Aristotle as much as the idea of empty time troubles us.

Aristotle's arguments against the void give us an interesting insight into his theory of motion. He invites us to consider the following imaginary experiment. Suppose a void to exist. How would an object placed in that void behave? Since there is nothing else in the void, there are no forces acting upon the object, so nothing obliging it to move in one direction rather than another.

This may strike us less as a damning indictment of the idea of a void as the derivation of an interesting consequence of that idea. But of course we would not accept that it is a genuine consequence. For one thing, there can be forces in a void, such as gravitational forces. Were it not so, the planets would not orbit the sun. In contrast, Aristotle's 'theory of gravitation', as we might put it, was that objects tend to move to their natural place, namely the centre of the cosmos, which for Aristotle is also the centre of the Earth. In a void, however, there is (he says) no up or down, nor any way of differentiating one direction from another. So if everything that exists does so in a void, as the atomists believed, then we would have no explanation of the motion of objects. Now the question that occurs to us, when Aristotle asks us how a body placed in a void would behave, is whether and in what direction the object was moving when it first entered, or was placed in the void. If it was not moving, then it will remain at rest; if it was moving at constant speed in a certain direction, then it will continue to do so, unless other forces act on it. If it was accelerating, then it will begin to decelerate—again, if no forces act on it. For we now reject Aristotle's assumption that it takes a constant force to keep a body in constant motion.

Aristotle has a second objection to the notion of void, which goes as follows: the speed and direction of an object is, in part, a function of the viscosity or 'texture' of the medium through which it is travelling. Thus my hand can move more swiftly through air than it can, with the same effort, move through water, as water offers more resistance. And in a medium of variable viscosity, the moving object will tend to take the line of least resistance. The less resistance a medium offers, the greater the speed of an object, when the force propelling that object stays the same. But a void can offer no resistance whatsoever, so an object propelled suddenly into a void will rush off at infinite speed in all directions.

Apart from the historical interest of these arguments, they illustrate an important philosophical method: start with a hypothesis you wish to defeat and show that it must be false by deriving absurd consequences from it. We will encounter another example of this method later, when we consider Kant's proof of the unreality of time. However, the fact remains that Aristotle's arguments against the void are weak, in that they rest on a mistaken physics.

Jars, Pumps, and Barometers

Fair-minded as ever, Aristotle enumerates the arguments put forward by defenders of the void. One of them in particular is rather more impressive than any of Aristotle's own arguments against. It concerns the result of an experiment in which a jar is filled with ashes. A second jar, exactly the same size as the first, is put beside it, but is left empty. Then both are filled to the top with water, the quantity of water poured into each being carefully measured. Remarkably, the jar containing the ashes will take as much water as the empty jar. What this suggests (although Aristotle does not spell it out) is that the water, although apparently continuous, has tiny gaps in it—numerous voids, in fact—into which the ashes can go. This is, of course, exactly what does happen when a substance is dissolved: its molecules enter the spaces between the water molecules. Although we do not know the precise composition of the ash Aristotle was referring to, it is likely that it was at least partially soluble.

Aristotle's response to the experiment is, again, a not very strong one. He tries to reduce the explanation in terms of gaps to absurdity. Although his remarks are not entirely clear, what he seems to be implying is that, if the water is capable of absorbing the ashes at any point (and he attributes this idea to the proponents of void), then it must be completely void, i.e. void at every point, but in that case there would be no water at all, but empty space, and that of course we know to be false. But why should anyone insist that the water can take in the ashes at *every* point? Surely it can do so only where there are gaps, although these will be so close together that they cannot be distinguished. To assume that the water is uniform is plainly to beg the question against the atomists, whose whole point is that water, and indeed any substance, is *not* uniform (i.e. continuous).

Despite the efforts of the early atomists, the long dominance of the Aristotelian ban of vacua in nature did not come to an end until the seventeenth century. Towards the end of his life, Galileo Gallilei (1564–1642) was working on a problem encountered by Florentine engineers who were constructing pumps to clear mines of water. It seemed that below a certain depth, water pumps using valves would not work: 32 feet seemed to be the maximum depth at which water could be removed by these pumps. The problem was taken up by Galileo's pupil, secretary and later successor in the chair of mathematics at Florence, Evangelista Torricelli (1608–47). In 1643 Torricelli constructed an apparatus that in effect reproduced in miniature the conditions hampering the water pumps. He took a glass tube just over a yard long and closed at one end, and filled it with mercury. He then inverted the tube, taking care that no air got in, and immersed the open end in a trough of mercury. He found that the vertical column of mercury did not extend to the top of the tube, but fell to a height of about 11 inches. At the top was a space apparently filled with air, except that it could not be air, since none had been introduced. He further discovered that the precise height of the column varied both with atmospheric conditions and the height above sea level the apparatus was situated. The column was rather lower, for example, at the top of a mountain, where the air was more rarefied than at sea level. Torricelli had, in fact, invented the first barometer. At the top of the tube was what is still referred to as a Torricellian vacuum. It is only a partial vacuum since it is impossible to eliminate the mercury vapour. What would Aristotle have made of these discoveries? The presence of mercury vapour perhaps suggests an answer: there is in fact no vacuum at all here, we can imagine

Aristotle arguing, but rather a very finely textured air. The irony is that Torricelli himself, like Galileo before him, persisted in the conviction that a vacuum was an impossibility.

Around this time, the mayor of Magdeburg, Otto von Guericke, devised a pump for removing air from a vessel, based on a principle similar to Galileo's water-pumps. Although it was only possible to achieve a very partial vacuum, von Guericke was able to demonstrate some impressive results of his new invention. In a famous experiment, conducted in Magdeburg in 1654, he joined two great copper hemispheres together and pumped the air out. The resulting force binding the two hemispheres together was so great that not even a team of horses pulling on each hemisphere could separate them. An illustration in von Guericke's *New Experiments,* published in 1677, depicts no less than sixteen horses engaged in this unusual activity. So, nineteen centuries after the writing of Aristotle's *Physics,* the defenders of the void triumphed. But another 200 years were to pass before the atomists could claim the status of orthodoxy for their theory.

Lessons of the Vacuum

The existence of vacua in space is an established fact. What philosophical conclusions can we draw from it? A number of reasons incline us to think that a spatial vacuum is less problematic than its temporal counterpart. These stem from the commonplace that, whereas time has only one dimension, space has three. The implications of this fact are that there is no plausible spatial analogue of any of the three arguments against temporal vacua we discussed in the previous chapter. Consider first the experience argument, based on the premiss that there can be no experience that we could recognize as an experience of a temporal vacuum. There is no spatial analogue of this, since we can have experience of a spatial vacuum, in the sense that we can look into it, and say 'Look, a vacuum.' Consider next the measure argument, based on the premiss that we cannot (directly) measure the duration of a temporal vacuum. In contrast, we can measure the size of a spatial vacuum, since we have access to its boundary, the dimensions of which enable us to compute the volume of space occupied (if that is the right word) by the vacuum. Consider finally the sufficient reason argument, based on the premiss that there can be no causal explanation of the ending of a temporal vacuum. In contrast, we can give an explanation of the boundaries of a vacuum: the vacuum is this big because of what exists beyond its boundary. We can, for example, give an explanation of the size of the vacuum at the top of a mercury column in terms of the weight of the mercury and the atmospheric pressure outside.

However, when we come to ask, could space exist if there were *no objects whatsoever* in the world?, the existence of spatial vacua is just as significant in helping us to settle the question as the possible existence of temporal vacua would be for our understanding of time. We have to be careful, however. Temporal vacua would establish that time could exist in the absence of change (obviously, since that is what a temporal vacuum is: time without change). But from the existence of a local spatial vacuum—a vacuum in only part of space—we cannot immediately jump to the conclusion that there could be a vacuum occupying the *whole* of space. Vacua, as we might put it, may need objects around them. To see this, we need to outline the debate between two theories of space.

The first theory is generally called *absolutism about space* (although it also goes by the name of substantivalism), and holds that space exists independently of objects: that is, the existence of space does not require there to be anything else in the world. Space contains objects, rather like a box, only in this case we suppose the box to have no sides. Space is, for the absolutist, an additional object in the world, something by virtue of which objects have spatial positions and stand in spatial relations to each other.

The rival theory, *relationism about space*, rejects the notion of space as a container, and the implication that space is an object in its own right. If there were no other objects in the world, according to the relationist, there would be no space. Space is not a thing, but rather a system of relations—spatial relations—between objects. Thus positions, or points, in space are defined according to their distances from objects. And this is not just a matter of convenience, or ease of identification; distances from objects is exactly what those positions are. Clearly, those relations would not exist on their own, with nothing to relate, hence the dependence of space on objects. Leibniz, who provided a number of arguments against absolutism, draws a useful analogy which likens space as the relationist sees it to a family, or genealogical tree:

[T]he Mind can fancy to itself an Order made up of Genealogical Lines, whose Bigness would consist only in the Number of Generations, wherein every Person would have his Place. (Alexander 1956, 70)

A family is simply a group of people standing in certain relations to each other: daughter, uncle, cousin, etc. The relations are something additional to the collection of individual people, but can we therefore talk of the family as an object in its own right, containing different people at different times? Leibniz develops the analogy further:

and if to this one should add the Fiction of a Metempsychosis, and bring in the same Human Souls again; the Persons in those Lines might change place; he who was a Father, or a Grandfather, might become a Son, or a Grand-son etc. (Ibid. 70–1)

Even without this particular fiction, we do talk in abstract terms of the family, as when we say, for example 'the nuclear family is under threat', or 'the Sixties changed the way we look at the family.' But we are using 'the family' here as a general expression, not as a means of referring to some particular item in the world. In similar terms, we talk of 'the Papacy', or 'the Presidency of the United States', not because there is some concrete item which persists through changes of holders of the office, but because different people at different times come to stand in the same relations to other people or institutions. As Leibniz goes on to say of his family analogy:

And yet those Genealogical Places, Lines, and Spaces, though they should express real Truths, would only be Ideal Things. (Ibid. 71)

Thus the family, as a concrete object, exists only if the constituent people do. Likewise, space exists only if its contents do.

Now relationism about space (unlike its temporal counterpart) must allow volumes of space that are not occupied by any object. Indeed, it would be in trouble if it did not allow these. But those empty places are, literally, nothing at all. There is a gap between two objects, say this chair and that window, not because there is some third entity between them, but because they are 5 yards apart and nothing is closer to either of them along the straight line from one to the other. Remove the objects and you are left with nothing.

The absolutist will reply that these distance relations only obtain between objects because of the existence of actual, unoccupied spaces. Again, drawing on the family analogy, but this time to illustrate the absolutist position, Tom is only Lucy's uncle by virtue of the fact that Lucy's mother is Tom's sister. Tom and Lucy do not stand in the family relation they do directly, but only by virtue of the relations they stand in to some third entity (Tom's sister). Their relation is, as one might put it, mediated. Likewise, for the absolutist, spatial relations are mediated. The relationist will object that these intervening spaces have no role to play. Why, they will say, should distance relations not be direct and unmediated? Here the absolutist might point to a number of roles these unoccupied points could play. First, they could play a role as objects of reference. 'Put the flowers over there,' I say, indicating an empty space in the corner of the room. What does the 'there' refer to? Something, presumably, but not to any perceptible object. I cannot be referring, for example, to an area of the carpet. I do not want the flowers to be put in the area presently occupied by that part of the carpet, requiring the flower-bearer to start digging a hole in the floor.

One could say that unoccupied places are possibilities of location, and that I am referring to those possibilities. But by virtue of what do genuine possibilities of location exist? Because, we surely want to say, there exist places in which things could be put. Here, then, is a second role for space: to make it possible for things to go where at present there is nothing (except space).

A third role for space is to account for the truth of certain geometrical statements. Consider the following statement: the midpoint on a line between the chair and the window is 5 feet from the end of the bookcase. For this to be true, presumably, there must be a midpoint. Moreover, since there is as a matter of fact no object between chair and window, it is an unoccupied point. So unoccupied points really exist.

What can relationists say in response to these putative roles for space? They will deny, first, that I am *referring* to anything when I say 'Put the flowers over there.' I am, rather, indicating roughly what distance relations I want the flowers to stand in to other objects. Secondly, they will say that what grounds the possibility of location, say between two objects, is not the existence of space, but rather the existence of spatial relations between those objects, together with the fact that there is nothing presently between those objects.

The third point, concerning the truth of certain assertions about unoccupied points between objects, is rather harder to dispose of. It raises an important general issue: what is the relationship between the abstract truths of mathematics and the world of concrete objects? As far as arithmetical truths are concerned (propositions such as $99/22 = 4\frac{1}{2}$), few would doubt that these are *necessarily* true: true no matter what the state of the physical universe is. Indeed, if the very existence of the physical universe is merely contingent (i.e. it might not have existed), then arithmetical truths remain true even in the absence of a universe. But the independence of mathematics from the world does not prevent it applying to the world: if you add two bananas to three oranges, you get five items of fruit. Now let us assume that what goes for arithmetic goes also for geometry. So, for example, if Pythagoras' theorem is true, then it is true whatever else is true of the physical universe. (In the next chapter, we will see that developments in nineteenth-century mathematics suggested a different treatment for arithmetic and geometry, and that geometrical truths may after all depend on contingent features of the world. But for the time being we shall treat the two branches of mathematics as equally indifferent to the world in which we live.) Now, given this view of mathematical truth, the relationist can deal with the midpoint problem as follows. Let us say that there are three objects situated thus: the chair is 10 feet from the window, and the end of the bookcase is 7 feet from both. We have here, then, three spatial distance relations. Now, it is a geometrical truth that if points A and B are 10 units apart, and point C 7 units from both, then the midpoint on a line between A and B is just under 5 units from C. This abstract piece of geometry does not require real points in space for its truth. By appeal to a combination of real spatial relations and abstract geometrical truths, then, the relationist can explain the truth of statements such as 'the midpoint on the line between chair and window is 5 feet from the end of the bookcase' without having to concede the actual existence of that midpoint. Given the actual arrangement of chair, window, and bookcase, it is pure geometry that guarantees that if an object is placed precisely between chair and window, it will be approximately 5 feet from the bookcase.

The Redundancy of Space

Space, says the absolutist, is capable of playing a number of roles: as something we refer to when we talk of positions ('over there', 'halfway'), as something that makes other things possible (as when an object could move into a currently unoccupied position), and as something that explains the truth of geometrical propositions. However, as we have seen, the relationist can simply deny that we need space to play these roles. And now the relationist goes on the offensive: not only does space as the absolutist sees it have no characteristic role to play, it is an embarassment, introducing supposed facts which we cannot explain. Earlier, I suggested that there is no plausible spatial counterpart to the sufficient reason argument against temporal vacua, but that is not quite true. I only considered the case of a vacuum being surrounded by the world of concrete objects, which thus define its boundary. But what if we consider the reverse of this: the world of concrete objects surrounded by a vacuum? Even to make sense of this, we have to subscribe to the absolutist theory of space (contrast the enclosed vacuum, granted by both absolutist and relationist theories). But if there is space beyond the boundaries of the universe—'extracosmic space'—then there are an infinite number of positions in space that the universe could occupy, compatibly with the present relations between the various components of the universe. In other words, we cannot tell, simply from observing the internal spatial relations of the universe *where the universe is* in relation to space itself. This in itself might be thought of as a problem, but there is another. We can put the point initially in theological terms. Since, as Leibniz puts it, 'God does nothing without reason', and no reason could be given for creating the universe in *this* part of space rather than *that* part—different parts of space being in themselves completely indiscernible—it follows that God cannot in reality have been faced with a choice between different places into which to put the newly created universe. The idea of empty space surrounding the universe, then, is a fiction.

There is a non-theological version of this argument that is worth considering, and it mirrors the sufficient reason argument against temporal vacua:

The argument against extracosmic space
  1. If it is possible for an object to occupy a position other than the one its does in fact occupy at a particular moment, then there is a causal explanation for why it occupies the position it does at that moment.

2. If the universe is surrounded by empty space, then it is possible for the universe as a whole to occupy a different position in space from the one it does currently occupy.

Therefore:

3. If the universe is surrounded by empty space, then there is a causal explanation for why it occupies the position it does.

4. But, since empty space is completely uniform (i.e. there is nothing to differentiate one region from another), there could be no explanation of why the universe as a whole occupies the position it does.

Therefore:

5. The universe is not surrounded by empty space.

Now the conclusion of this argument is not actually inconsistent with absolutism, since there are two possibilities left: (i) The boundaries of absolute space coincide precisely with those of the finite universe, so there is no possibility of the universe's being situated elsewhere in absolute space. However, this option would at the least be somewhat embarrassing for the absolutist, for the coincidence of the boundaries of space and universe would be just that, a *coincidence*: not something we should have reason to expect, since space is (for the absolutist) completely independent of the universe it contains. (ii) The universe, and therefore space, is infinite in extent. Here, there is no need to explain the coincidence between the boundaries of space and the universe, for there would be no boundaries.

The absolutist could also query the inference from 1 and 2 to 3. For it could be argued that premiss 1 is true only of objects *in* the universe; it is a moot point whether it can plausibly be extended to the universe as a whole. For the position of objects is affected by the behaviour of other objects. The position of the universe as a whole, however, cannot be affected by any other object, since there is no other object (with the possible exception of God), merely empty space. So perhaps the demand for explanation is simply inappropriate.

We seem to have reached an impasse. These arguments concerning empty space, though suggestive, do not seem able to settle the debate between the absolutist and the relationist. It is time to look elsewhere.

The Search for Absolute Motion

The original debate between relationist and absolutist views of space arose from the Newtonian theory of motion. Newton had distinguished between two kinds of motion: relative and absolute. *Relative motion* is the change in spatial relations between objects. Imagine two people in conversation as they descend into the depths of the London Underground. One descends by means of the stairs, the other on the escalator. Relative to the walls around them, they are moving at the same rate. As a consequence (since they are moving in the same direction), both are stationary relative to each other, and so can continue their conversation. But now the one descending the stairs stops to tie his shoelaces. His companion, to be polite, now starts to ascend the downwards-moving escalator, so that she remains stationary relative to her friend, and as a result is moving with considerable speed relative to the escalator steps. Now, in addition to these relative motions, there is, or so we are powerfully inclined to suppose, what we might call *real* motion. If A is in motion relative to B, then B is correspondingly in motion relative to A. But is it A or B that is really moving, the other remaining stationary? Or are they both moving? These, at any rate, are the questions we want to ask. When Gallileo first asserted that it was the Earth that orbited the Sun, and not the other way around, that was a heretical and dangerous position to hold, and he was forced to recant by the Spanish Inquisitors. Why did he not rest content with the assertion that they were in relative motion? Of course, it may be more convenient to say that the Earth orbits the Sun, rather than vice versa, since the Sun is stationary in relation to the stars, and the Earth is not. But is it simply a matter of convenience? Is the Sun not really at rest, not merely in relation to the stars, but absolutely?

Absolute motion is the name given to motion relative to space itself. If there are such things as points in space, independently of any other object, then these points are by definition stationary (since to be stationary is to stay in the same place and a point *is* a place). Anything which moves relative to these points is in motion. The notions of absolute motion and absolute space—that is, space as the absolutist conceives of it—belong inseparably together. So any argument for absolute motion, any argument which purports to establish that we cannot make do with relative motion alone, is an argument for the absolutist theory of space.

We have already encountered one argument against absolute space: the impossibility of explaining (or indeed knowing) where the universe is in absolute space. Another argument in the same spirit is Leibniz's argument from the undetectability of absolute motion:

To say that God can cause the whole universe to move forward in a right line, or in any other line, without making otherwise any alteration in it, is another chimerical supposition. For, two states indiscernible from each other, are the same state; and consequently, 'tis a change without any change. (Alexander 1956, 38)

Just as we cannot say where the universe is in absolute space, we cannot tell whether the universe as a whole is stationary in that space or moving through it with constant velocity. Similarly, when we are travelling in our car at constant speed down a road, there is nothing *within the car* to distinguish our condition from that of rest with the engine running (apart from the effects of bumps or swerves in the road). So motion which is *only* motion relative to space, which therefore involves no change in the spatial relations between objects, and which is of constant velocity in an unchanging direction, is indistinguishable from rest. Such motion, concludes Leibniz, is 'chimerical', by which he could mean any of the following: that it is actually an incoherent notion, that we have no clear idea of it, or that, although coherent, it could play no role in any explanation. To show that absolute motion was actually incoherent, we would need to appeal to some version of the verificationist account of meaning familiar from previous chapters. But the relationist need not go to such extremes. It is enough of a recommendation for relationism, surely, if it turns out that absolutism is an unnecessarily extravagant theory, introducing something that explains nothing.

But, as Leibniz clearly recognized, the debate does not end there. It may be true that we do not notice the effects of constant motion, except in terms of changing position vis-à-vis other objects. But we *do* notice the effects of *in*constant motion, of changing velocity or direction, even in the absence of signs of relative motion. Accelerating down the runway in an aeroplane, we are forced back into our seats. Braking sharply in a car, we are flung forward. Swinging a heavy object on a piece of rope around our heads, we feel the tug as it pulls away from us. So if the universe as a whole were accelerating through absolute space, then we would, presumably be aware of the motion. It is *forces*, then, which may be the key in our search for absolute motion, and which therefore provide evidence for the existence of absolute space.

In his massive and hugely influential *Principia Mathematica*, Isaac Newton (1642–1727) imagined two simple physical set-ups to illustrate the connection between forces and absolute motion. One involved two identical globes, connected to each other by a cord. He then invited us to imagine the globes rotating about the midpoint on the cord. The centrifugal forces which resulted from the motion would produce a detectable tension in the cord, the magnitude of which would vary with the speed of rotation. Now the significance of this imagined experiment lies in the fact that the globes are not in motion relative to each other: the distance and orientation of each globe vis-à-vis the other never alters. From the perspective of each, the other is stationary, though both may be in motion with respect to other objects. But now suppose further that these globes are the *only* objects in the universe, so that there is no relative motion at all. Tension in the cord would then indicate that the globes were in absolute motion.

The second experiment (Figure 1) involved a bucket full of water suspended from a cord, allowing the bucket to rotate about its vertical axis. The experiment has four stages: (i) bucket and water are initially at rest; (ii) the bucket is made to rotate, but because of inertia the motion is not immediately transferred to the water, so water and bucket are in motion relative to each other; (iii) as inertia is overcome, the water begins to rotate so that bucket and water are now approximately at rest relative to each other, but the centrifugal forces thus generated create a concavity in the surface of the water; (iv) the bucket is made to stop rotating but, again because of inertia, the water continues to rotate for a time and the surface remains concave; bucket and water are once again in relative motion. Now, what is the difference between (ii) and (iv)? (Newton actually describes only the first three stages, but because water and bucket are not completely at rest relative to each other in (iii), it is best to compare (ii) and (iv) rather than (i) and (iii).) As far the relative motion of bucket and water is concerned, nothing at all. Why then is the water surface flat in (ii) and concave in (iv)? The answer it seems is that the water is absolutely at rest in (ii) but absolutely in motion in (iv). The relationist could point out that the water is at rest relative to objects without the bucket in (ii) but in motion relative to those same objects in (iv), but as with globes, we can imagine there to be no other objects outside the experiment. The surface of the water would then indicate absolute motion or absolute rest.

Whether or not it was Newton's intention to present these experiments as establishing the existence of absolute space, or simply as illustrating the explanatory value of such a notion, we can use them to construct a plausible argument for absolute space. Abstracting the essential principles from the details of the two experiments, we have the following:

(i)
Bucket and water
at rest.
Surface of water
flat.

(ii)
Bucket starts to move,
but movement not
yet communicated to
water. Surface of
water flat.

(iii)
Bucket and water in
motion, but at rest
relative to each other.
Surface of water
concave.

(iv)
Bucket at rest, but
water continues to
be in motion.
Surface of water
concave.

Fig. 1. The bucket experiment

The argument from absolute acceleration

1. The presence of centrifugal forces in a system whose components are at rest relative to each other is attributable to acceleration.

2. An object's acceleration is either absolute—i.e. relative to absolute space—or merely relative to other objects.

3. It is possible for centrifugal forces to be present, within a system whose components are at rest relative to each other, in the absence of any objects outside the system, and so in the absence of any relative acceleration.

Therefore:

4. It is possible for there to be absolute acceleration.

Therefore:

5. Absolute space exists.

The move from 4 to 5 involves the following line of thought: given the definition of absolute acceleration, namely as acceleration relative to absolute space, acceleration is only possible if absolute space exists.

The premiss which requires a leap of faith (since it is neither a matter of definition nor something we can experimentally test) is 3. What entitles us to assume that objects without the system are irrelevant? Perhaps in the absence of such objects, no system would exhibit the force effects normally attributable to acceleration. On the other hand, the force effects observed in ordinary circumstances appear to be entirely internal to the system. To the extent that other objects impinge (which of course they do to some extent, for example by exerting gravitational forces) it is hard to see how this limited interference makes a significant contribution to the characteristic centrifugal effects of rotating systems. Alternatively, we could allow that force effects might be present in an isolated system, and allow further that such effects are ordinarily observed in accelerating systems, and yet find no sense in the suggestion that the system exhibited any acceleration amongst its components, precisely because one could only make sense of relative acceleration.

We might just pause here to note how we have been using the terms 'relative motion' and 'relative acceleration'. We have used them to refer just to the motion of an object relative to *other* objects. But why could we not use the terms also to refer to the motion of an object relative to *itself at earlier times*? Of course, at every instant, an object is at zero distance from itself at that same instant, but it is not necessarily at zero distance from itself at *other* instants. So one and the same object at different instants may be considered in certain contexts equivalent, or at least analogous, to different objects. So perhaps the relationist *can* make sense of a universe consisting of a single object in motion, and therefore of an accelerating system of objects not in motion with respect to each other. Tempting though this thought may be, however, it would not be wise for the relationist to extend the notion of relative motion in this way. For a relationist about space is likely also to be a relationist about time (given the similarity of some of the considerations in favour of each). Now, if we are to consider an object as being in motion relative to itself at earlier times, this presupposes that we have some way of individuating times—i.e. an account of what makes different times different. The relationist about time will say that times are different by virtue of the fact that different states of affairs obtain in them—indeed, times just *are* states of affairs. How, then, in a universe with a single moving object, are earlier times to be distinguished from later ones? In terms of the different positions of the object, of course. But how did we distinguish these different positions? In terms of spatial relations between the object and itself at different times. So, in defining position, we presupposed times, and in defining times we presupposed position, thus arguing in a circle. Relative motion, then, had better mean just motion relative to other objects.

The encounter between absolutist and relationist has so far been inconclusive. The absolutist cannot force the relationist to accept all the premises of the argument from absolute acceleration. Nevertheless, the absolutist appears to have gained some ground. For the traditional, Leibnizian response to absolutism is that, *even on absolutism's own terms,* space plays no important explanatory role. Neither absolute position nor absolute motion makes any difference to what we can observe. But Newton's globe and bucket experiments seem to show that absolute space is not necessarily idle: it is the absolute acceleration of objects through space that explains the forces on them. Unfortunately, even this is too strong a conclusion to draw at this stage. The relationist can insist that it gets the order of explanation the wrong way around. It is force which is the cause of motion, not the reverse. What requires explanation, therefore, is not the existence of forces, but the effects of those forces, namely motion. Absolute space itself is inert, and it is inert because it is homogeneous—everywhere the same.

Or is it? In the next chapter, we shall consider the suggestion that space may have some rather surprising properties, properties that provide the missing element in the absolutist's attempt to show that space is an explanatory item in its own right.

Questions

If everything is composed of atoms, must there be empty space between them?

Aristotle said that 'Nature abhors a vacuum.' How then would he have accounted for the outcome of von Guericke's Magdeburg experiment?

What do you make of the following remark: 'Space must be an object in its own right. For if it were *nothing*, there could never be empty spaces between things'?

CHAPTER 4

Curves and Dimensions

As lines, so loves, oblique may well
Themselves in every angle greet:
But ours so truly parallel,
Though infinite, can never meet.

Andrew Marvell, 'The Definition of Love'

Euclid Displaced

We begin with a puzzle. Imagine you are sitting inside a room with no window and whose only door is both shut and locked. There are no holes in the walls, floor, or ceiling. In what circumstances would you be able to leave that room without either opening the door, digging a tunnel, or knocking a hole in the ceiling or any of the walls?

To see a way out of this uncompromising situation, we need to rethink some of our ordinary ideas about space. Let us first consider straight lines and their properties. A straight line we will define as that which marks the shortest distance between two points. How many straight lines could join two points (taking a point to be dimensionless—i.e. not taking up space)? The natural answer is that since, of all the routes we could take from one point to another, there is only one shortest route, it must be that there is one unique straight line through any two points. Consider, then, two straight lines drawn on a flat surface. Suppose the two lines to be extended infinitely in both directions. Do they meet at any point? If they do, then we say the lines are *oblique*. If they do not meet at any point along their infinite lengths, then we say that the lines are *parallel*. Now consider one of those lines, and a point not on that line but somewhere on the surface. How many lines parallel to that line can we draw through that point? A few rough sketches incline us to the natural answer: one and only one. So now we have two principles concerning lines on a flat surface:

(A) Precisely one straight line joins any two points.

(B) Through a point not on a given straight line, there is precisely one straight line parallel to the first.

These two principles are fundamental truths according to the geometry systematized by the Greek mathematician Euclid, who taught at Alexandria in the third century BC. At the beginning of his monumental work, *The Elements*, he lays down definitions of such notions as point, line, and angle, followed by a statement of fundamental, self-evident truths, which form the starting point of his system. He then proceeds, in the thirteen books of *The Elements*, to prove, from the fundamental postulates and definitions, by means of rigorous methods of deduction, several hundred theorems—propositions that, though not themselves self-evident, can be derived from self-evident propositions. Among these theorems is the proposition that the internal angles of a triangle add up to 180. The fact that Euclid's theorems are derived from propositions taken to be self-evidently true conferred on them a special status: they were taken to be a set of logically necessary truths, which could not fail to be true. Anyone crazy enough to deny them would be guilty of self-contradiction. In this, they are seen as quite different from, say, a list of biological facts, things which, though as a matter of fact true, could have been otherwise.

This peculiar status of Euclid's geometry as the One True Geometry went unchallenged until the end of the nineteenth century. There had for some time been worries about principle (B) (or, to be precise, an essentially equivalent principle, but the details of Euclid's formulation need not worry us). It was listed as one of Euclid's fundamental postulates but, for some reason that is hard to discern, it was felt to be rather less self-evident than Euclid's other postulates. (B), it was felt, should be a theorem, and so capable of demonstration. In the eighteenth century, an Italian priest named Girolami Saccheri set out to find this proof. What he hoped to show was that a denial of (B) would result in contradiction, as one would expect of a necessary truth. He failed to do so, but felt sufficiently confident that he had put (B) on a firmer footing by deriving some repugnant conclusions from its denial, that he published his results in 1733. Saccheri's project was taken up by Carl Gauss, later famous for his work on electro-magnetism, and it seems that as early as the 1790s, he had conceived the idea of a geometry in which (B) was not true. Quite independently of Gauss, a Hungarian mathematician, Janos Bolyai, developed during the 1820s a theory of parallel lines that contradicted Euclid's postulate. Bolyai's father showed his son's work to Gauss, with whom he had been at university in Göttingen. Gauss is said to have remarked that he had been thinking along the same lines for years, but, feeling rather insecure about his work, had never felt able to publish the results. The first person to publish a description of what became known as non-Euclidean geometry was the Russian Nikolai Lobachevski in 1829, although his work was initially very poorly received.

What exactly were Gauss, Bolyai, and Lobachevsky trying to do? Euclid's principles describe the properties of lines and shapes on a flat surface. But what if the surface is not flat? Then lines will behave rather differently. On a flat surface, a straight line can be extended indefinitely in both directions. But consider what happens to a line on the surface of a cylinder (Figure 2). This line cannot be extended indefinitely in both directions, since it will eventually come back on itself. Someone will object that this is not really a straight line at all, by virtue of the fact that it is not on a flat surface. But if you consider two points, *a* and *b*, also on the surface of the cylinder, and through which our line goes, there is no shorter route on the surface between these points than that very line. So although we perceive the line as curved, it nevertheless satisfies the definition of a straight line, since on the surface described, the line *is* the shortest distance between the two points.

The next significant development came in 1854, when the young German mathematician Georg Riemann submitted to the University of Göttingen an application for the post of *Privatdozent* (unsalaried lecturer) in the form of a dissertation, whose theme had been set by Gauss himself. The essay developed the idea of a *spherical* geometry, in which postulates (A) and (B) are false. To see how, imagine the surface of a sphere in which a number of straight lines have been drawn (Figure 3). As with the cylinder, the shortest distance between any two points will appear to us as a curve. Lines on the cylinder, however, satisfied both (A) and (B). But here there are no straight lines that do not meet at some point. In Figure 3, all lines meet at the poles. So

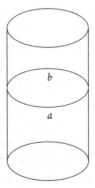

Fig. 2. A 'straight' line on a cylinder

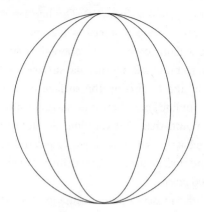

Fig. 3. Parallel lines on a sphere

Principle (B) fails. Principle (A) also fails, since the points at the poles are joined, not just by one straight line, but an infinite number of them.

So far, we may seem to have done very little to unseat Euclid from his place as the geometrical lawgiver. For Figures 2 and 3 are representations of three-dimensional objects, and we all know that in order to find the truly shortest distances between points on the surfaces of these objects, we would have to pass a line above or below the surface. So these examples do not really challenge Euclid's postulates. Or so it may be argued. But this line of reasoning is mistaken. Accustomed to interpreting certain two-dimensional figures as representing depth, we naturally see Figures 2 and 3 as depicting three-dimensional objects. But let us now suppose that what they in fact represent are spaces of only two dimensions: spaces in which the inhabitants may move North, South, East, and West, but not up or down, for there is no up or down. It would be natural to think of a two-dimensional world as completely flat (indeed when in 1884 E. A. Abbott published a book concerning the goings-on in a two-dimensional world, he called it *Flatland*), but we are not obliged to do so. We can suppose instead that a two-dimensional world is shaped like the surface of a sphere. It is not in fact the surface of a sphere, for a sphere requires three dimensions. However, the points in this imaginary two-dimensional world are related to each other exactly as are the points on the surface of a sphere (where we confine our attention to relations in two dimensions). Clearly, Euclid's geometry will not adequately describe such a world. Of course, if the world is a sufficiently large one, then things will seem to be Euclidean for its inhabitants, just as a small part of the surface of a very large sphere looks pretty flat. But it ought nevertheless to be possible, by taking appropriate measurements, to discover whether the two-dimensional world is Euclidean or not.

The moral of this thought experiment is that no one geometry is true as a matter of logical necessity: geometries that differ from Euclid's are just as consistent as his. And what goes for the two-dimensional world we have been describing goes also for three-dimensional worlds. So it is an open question which of a variety of consistent geometrical systems correctly describes the space we inhabit. Our three-dimensional space may be a curved space, and its curvature may be constant or variable. Space, as we might put it, is something that has a *shape*.

Space Makes Its Presence Felt

Having seen how pure mathematics alone does not determine the properties of space, we now revisit some of the arguments for and against absolutism about space that were introduced in the previous chapter, to see how the possibility of non-standard geometries makes a difference to the force of those arguments.

One role for space the absolutist identified was to account for the truth of certain mathematical statements concerning distances. A, B, and C are occupied positions, but given information about the distances between them, we can make assertions about unoccupied points, such as 'The midpoint on the line from A to B is n units from C.' The midpoint does not have to be occupied itself for this statement to be true. But the truth of statements apparently about unoccupied places seems to require the actual existence of such places, over and above anything that might occupy them. This, surely, is something the relationist wishes to deny. A possible relationist strategy, it was suggested, was to treat geometrical statements as not, strictly speaking, about the world at all, but about abstract relations in the world of numbers and ideal shapes. The midpoint on the line AB is *a purely mathematical*, not a physical notion, it was suggested. However, as we have just seen, this abstract world does not contain a single, true geometry. There are a number of geometries, descriptive of different spaces. That means that the relationist cannot appeal to the eternal truths of mathematics to account for the truth of distance statements concerning unoccupied points. In our earlier example, the value of the distance between an occupied point, C, and an unoccupied point, the midpoint on AB, can be calculated by means of Pythagoras' theorem. But that theorem is false in some non-Euclidean geometries. So the truth of any such distance statement must be accounted for in terms of the actual physical state of affairs that obtains, and that state of affairs must involve the existence of unoccupied points.

Consider another argument for absolutism, the argument from absolute acceleration. The absolutist argues that an object can be in motion, without necessarily being in motion relative to other objects, and that absolute acceleration (i.e. acceleration with respect to space itself) can explain what would otherwise be mysterious forces. The relationist, however, will either deny that forces would be present in the absence of relative motion, or deny that, if they were present, they would require explanation. And so we reach stalemate. But the discussion of non-Euclidean geometry in the last section now gives the absolutist a further explanatory tool. One moral of that earlier discussion was that, although space may be entirely uniform, it also may be variably curved, or indeed have any of a number of shapes. This curvature, or lack of it, constrains the motion of objects through space. Two objects moving initially along parallel paths may find themselves converging, not because some force has been brought to bear on them, but simply because their paths take them through a curved part of space. (Think of two ladybirds climbing up opposite sides of a bottle. As they reach the top, they move nearer each other, but not because they are being *propelled* towards each other.) Now an object will be affected by its motion through space, whether or not that motion is constant or accelerating. If the space through which it is moving is curved, then some parts of the object may be compressed, others may be extended, since different parts of the object are traversing lines in space that, because of the shape of space itself, may converge or diverge at various points. The result is forces within the object. If there is no curvature, then there will be no resulting forces, but then the absence of forces calls for explanation no less than their presence.

Here is another illustration of the effect space could have on its objects. An argument against absolute space that has distinctly Leibnizian overtones was suggested by the French mathematician and philosopher of science Henri Poincaré (1854–1912). (His younger cousin, Raymond Poincaré, was President of France during the First World War.) Poincaré invites us to imagine everything in the universe doubling in size overnight. For the absolutist, this is a perfectly coherent possibility, for it involves a genuine change in the relations between objects and absolute space. But, asks Poincaré, would we *notice* such a change in size? No, clearly not, since we can only compare the size of things against other things, and the point of the thought experiment is that relative size has not changed at all. So the status of a sudden doubling is rather like that of a five-million-year temporal vacuum intervening between your reading this sentence and your reading the next: postulating its existence, or even mere possibility, seems otiose. Now, an assumption of this argument seems to be that the change in size will not result in a change in shape. But in the non-Euclidean geometries we have considered, a change in size *does* involve a change in shape. Consider a small triangle on the surface of a large sphere. If the triangle is small enough and the sphere large enough, then the sum of the internal angles will be very nearly, though very slightly greater than, 180. But if this triangle is now massively expanded, the gap between the sum of its internal angles and 180 will be much larger: its shape, in other words, will have changed. Again, if we think of the triangle as a physical object, this will induce internal forces, so the change in size really does have effects.

We might just pause to note that the relationist will not necessarily reject the idea of doubling in size, as it involves a real change in the distance relations between parts of the object. Only a relationist who was also a *conventionalist about spatial metric*—that is, one who thought that distance was only definable in terms of some physical measure would reject the idea of doubling, for if (to speak for a moment in absolutist terms) everything doubles, then the measuring rods also double.

Space, then, can be thought of as having a feature that explains the behaviour of objects moving through it, that feature being its shape. This certainly defeats the Leibnizian point that, even if we grant the reality of space as the absolutist conceives of it, it can play no explanatory role. Space can make its presence felt in determining the motion of objects, but not by exerting forces. This means that the absolutist needs to attach an important qualification to Newton's Laws of Motion: they may be appropriate for Euclidean space, but not necessarily for other spaces.

How will the relationist respond to these considerations? Consider first the argument about statements concerning unoccupied points. There seem to be three possible strategies the relationist could adopt, two of which are not particularly attractive. The first is to insist that, after all, there is only one true geometry, whether Euclid's or some other system. The difficulty this poses is that of accounting for the apparent consistency of logically distinct geometries. The second strategy is to allow that some perfectly meaningful assertions are neither true nor false. Since there are no unoccupied points to make true statements apparently about them, these statements simply lack a truth-value. Their truth-value is, as we might put it, *underdetermined* by the physical facts. This, again, is not an attractive position. Once we actually attach an object, such as a length of string, or a rod, to A and B, thus connecting them by a straight line, we can easily discover the distance between C and the midpoint on this line, and it is very hard to shift the intuition that there was a truth of the matter about this distance *before* we connected A and B.

The third strategy is to say that, after all, these apparently unoccupied points are occupied, not by ordinary physical objects, but by the regions, or 'fields', of gravitational and electrical force that surround such objects. These force fields exist even in a vacuum, which is why, the relationist will point out, we cannot legitimately infer, from the possibility of local vacua, the possibility that the whole of space might have been one giant vacuum. This strategy is essentially the same as the one suggested for the relationist about time at the end of Chapter 2: reduce times to states of affairs. Again, it seems to provide a compromise between the relationist and absolutist. It concedes to the absolutist that we cannot do without talk of points unoccupied by any object, but concedes to the relationist that space can be reduced to something more fundamental. The kind of relationism we are considering, then, simply substitutes force fields for the absolutist's space. It is the shape of these force fields that is described by physical geometry. This gives the relationist a way of explaining the behaviour of objects as they move through space. Any changes in their internal forces will be explained by distortions in the force fields through which they are moving. Of course, the absolutists will say that their view goes to a deeper explanatory level: the distortions, or lack of them, in the force fields are not just brute facts, but explicable in terms of the shape, the geometry, of the containing space. The relationist will retort that this is no explanation at all.

Has the absolutist any further weapon? There is one, and one that appears to be quite independent of motion and force: *chirality*.

The Lone Hand

I look into the mirror hanging on my sitting-room wall, and see a room that is both familiar and strange. I recognize each object: the piano, the sofa, the doorway, the fireplace. Their shape, size, and distance from each other are unaltered. Why then do they seem so different? The obvious answer is that their relations to each other have been reversed. What in the actual room is on the left of the window as I face it is, in the looking-glass room, on the right. But there is more to it than that, for as we look more closely, each individual object has subtly changed its appearance. The name of the maker on the piano has been transformed into Cyrillic; a picture that seemed well-proportioned now seems distinctly lop-sided; the statuette of a discus-thrower looks as if it is about to keel over. The asymmetry of things appears exaggerated. It seems that, after all, the mirror has altered the shapes, at least of some things, but how exactly?

In the 1760s a then rather obscure lecturer at the University of Königsberg was considering the problem of the difference between a hand and its mirror-image. His post, which he had held since 1755, was unsalaried, and he relied for his income on private tuition. His fortunes changed in 1770, however, when he was at last elected to the Chair of Logic and Metaphysics at Königsberg, and he remains an inspiring example of someone who produces their best and most influential intellectual work in later life. The work for which he is best known was published when he was 57, and although not particularly young when elected to the chair, he held it for 27 years. His name was Immanuel Kant (1724–1804), and his first remarkable thoughts on what the human hand could tell us about the nature of space itself were published in 1768, in an essay entitled 'Concerning the Ultimate Foundation of the Differentiation of Regions in Space'.

In that essay, Kant invites us to imagine something very strange indeed: a universe that is completely empty except for a single human hand, not connected to any body. Now this hand can be imagined to be either a right hand or a left hand—indeed, if it existed, it would have to be one or the other—but what determines which it is? Were it attached to a body, the matter could be settled at once, since the left hand is quite differently related to the body than is the right hand. But we have no body to help us (and we must not imagine ourselves situated in this universe). Is it then something *internal to the hand* that determines whether it is left or right? This cannot be right either. For suppose we take an exhaustive set of measurements of our right hand: distance from the tip of each finger to the wrist, width of palm, angle of thumb, etc., etc. All these correspond precisely to the properties of the 'left' hand we see when we hold up our right hand to the mirror. The internal spatial relations of the hand, then, are preserved in reflection. (Of course, there will be intrinsic differences between our two hands, but these have nothing to do with the fact that one is right and the other left.) But if it is not relations to other objects that determine the handedness of Kant's lone hand (for there are no other objects), nor the spatial relations between the different parts of the hand, then there is only one possibility left: the hand's relation to *space itself*.

The phenomenon Kant was interested in is not unique to hands: it belongs to any object that has no plane of symmetry. Such asymmetric figures cannot be made to coincide with (fit exactly the same space as) their mirror-image counterparts by any series of rigid motions (i.e. motions which preserve the shape of the object). A corkscrew is one such asymmetric object. The molecules of certain substances, such as adrenaline and nicotine, are also asymmetric, and occur in 'left-handed' and 'right-handed forms', and these different forms have markedly different physiological effects. These objects and their mirror-images are called 'incongruent counterparts' of each other, an expression which calls attention both to their similarity and also their distinctness. The property of not being superposable on their mirror-image is called *chirality* (from the Greek *kheir*, meaning hand). We may be tempted into thinking that asymmetry and chirality are one and the same property, but this would be a mistake, as we shall see later. 'Handedness' we will reserve to name the property of being either right-handed or left-handed: it is whatever it is by virtue of which a hand is, e.g. a right hand. Kant sees handedness as posing a challenge to the relationist about space. Because relationism explains the spatial properties of things entirely in terms of spatial relations, whether between one object and another, or between different parts of one and the same object, it cannot account for the difference between a universe containing only a left hand and one containing only a right hand. All the spatial relations that would be countenanced by the relationist are the same in both universes. Yet, there is a difference, and only the absolutist, who posits an extra entity—space itself—can explain it.

Kant was later to have a quite different view of space, not as an entity existing in its own right and independently of any mind but as a way of interpreting our experience so as to make it intelligible. (His reasons for doing so will be looked at in Chapter 6.) But it is his earlier view that concerns us here. Before proceeding, it would be helpful to present the argument more formally.

The argument from handedness

1. There is an objective difference between a universe containing only a left hand and one containing a right hand, namely the *handedness* of the hand in question.

2. Handedness is a spatial property.

3. The spatial properties of an object are determined by either (a) relations between that object and other objects, or (b) relations between the parts of the object, or (c) relations between the object and absolute space.

4. In the two universes, handedness cannot be explained by (a), as there are no objects besides the hand.

5. It cannot be explained by (b), as these relations are the same in the two universes.

Therefore:

6. It must be explained by (c).

Therefore:

7. Handedness requires the existence of absolute space.

The first point to note about this intriguing argument is the obscurity of the idea that handedness depends on, is somehow determined by, a relation between the hand and space itself. How, exactly, does space determine handedness? The obvious kinds of case in which spatial properties would depend on space itself involve the notion of absolute location, examined in the previous chapter. That kind of property would not be countenanced by the relationist, of course, and in any case it does not help us here since handedness is independent of location (it is preserved by rigid motion). Perhaps, then, Kant thought that the left hand and right hand could be distinguished by their different relations to points of space. But this cannot be right, for the relations between any part of the hand and any point or points outside the hand are preserved by reflection. In any case, this could not have been Kant's thinking, since he says that what determines handedness is the relation of the hand to space *as a whole*. It is possible, of course, that Kant himself did not have a clear idea of what exactly it was about space that determined handedness. He just knew that something did, and he had eliminated the other suspects.

Looking at the argument through the eyes of the relationist, two premisses are vulnerable. One is (1). Recall the argument from absolute acceleration. There we were invited to consider a thought experiment in which objects were subject to centrifugal forces normally associated with acceleration even though they were not moving with respect to any other object. Here we are invited to consider a hand exhibiting a determinate handedness in the absence of other objects. In both cases, the relationist seems entitled to say that we do not know exactly what properties objects would possess in total isolation. Perhaps the lone hand just does not have a determinate handedness at all: it is not definitely a left hand or a right hand. But here the relationist is up against strong intuitions. What would it be for a hand to be *neither* left nor right, despite being a hand, and having the characteristic asymmetry of hands? (We return to this question shortly.)

The second vulnerable premiss is (5), and here the relationist may have a more plausible means of attack, this time resting on the very intuitions which threatened the first manoeuvre. (5) is true provided that we understand 'relations between parts of the object' to be simply relations of distance and angle. But why must we accept that these are the only spatial relations there are? Perhaps handedness depends on some special kind of relation between the parts of the hand, not explicable in terms of distance and angle. The relationist, then, could insist that handedness is an intrinsic, fundamental property of objects: intrinsic, in that it is independent of the properties of any other object, and fundamental, in that it is not explicable in terms of other properties. This move does fit with our intuitions and experience. Just looking at our hands shows us their distinctness: they do not look the same shape (not even the same shape viewed from two different angles). We do not have to see them in the context of other objects to tell which is which. So it is a very plausible supposition that handedness is intrinsic, and that (5) therefore is false.

Plausible, but wrong. Handedness is *not* an intrinsic property of the hand, despite appearances to the contrary.

More than Three Dimensions?

To see this clearly, it helps to simplify matters by considering just two dimensions. In Figure 4, we have two shapes that are mirror images of each other. Provided that we confine them to the flat plane of the paper, they remain incongruent: we cannot make them coincide however we move them around, not, at any rate,

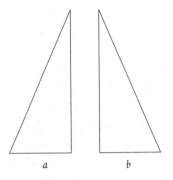

Fig. 4. **Incongruent counterparts**

without distorting their shape. Call *a*, the shape on the left-hand side, A-shaped, and *b*, the shape on the right-hand side, B-shaped. Now they certainly *look* as if they are different shapes. But are they? No: they only seem different shapes because of the way they have been arranged before us. Let us now allow them to move in the third dimension, perpendicular to the surface of the paper. Now *a* can be made to coincide with *b*, just through our turning it over. So the two objects are not, after all, different shapes. The difference between being A-shaped and being B-shaped is just a matter of perspective, not a difference in the intrinsic properties of the objects.

Consider again our hands. They are just the three-dimensional counterparts of the two-dimensional shapes in our Figure 4. Like *a* and *b*, they look different in shape. But they are not. There is no more intrinsic difference in shape between being left-handed and being right-handed than there is between being A-shaped and being B-shaped. The apparent difference, as before, is just a matter of perspective. It is true that, as long as they remain confined to three dimensions of space, they are incongruent. But now suppose we introduce a fourth dimension. Then we should be able to make the two hands coincide, just as we made *a* and *b* coincide by rotating one in three dimensions. This is, of course, hard for us to appreciate, since we can only see things in three dimensions, and this constrains what it is possible to imagine. And since, arguably, we develop the concept of three dimensions through perception we find it hard to conceive of what it is for space to have more than three dimensions. In this, we might be assisted by a definition of dimension, an idea we so far have been taking for granted. A *dimension*, in the widest sense, is simply a way in which a thing can vary with respect to a given property. Thus we can talk of the dimension of pitch, intensity, hue, or temperature. A given space has n dimensions (where n stands in place of any given number) if there are n independent ways in which an object can be spatially distant from another object or location. So, for example, an object may be 3 yards north of me and 2 yards above the level of my feet. Now I can vary the extent to which the object is north of me without having to alter the extent to which the object is above the level of my feet. We assume that there are precisely three independent ways in which two things can be spatially distant from each other. But there is no logical inconsistency in supposing there to be a fourth spatial dimension: a fourth independent way in which something can be distant from something else. After all, although Euclid's geometry confines space to three dimensions, we have seen that Euclid's is not the only possible geometry.

Although this means that the relationist can no longer defeat the argument from handedness by denying (5), it strengthens the other strategy, namely denying (1). If the difference between left-handedness and right-handedness is not intrinsic, but just a matter of perspective, then a lone hand is not determinately left or right. The counterintuitive strategy turns out to be the right one, and the argument from handedness fails. But that is not yet the end of the story. For although the handedness of the lone hand is not determinate, something else is: whether or not it is *chiral*, i.e. can or cannot be superposed onto its mirror-image. The A and B shapes were chiral in two dimensions, but lost their chirality once they were free to move in three dimensions. And so it is with hands. If confined—as we suppose them actually to be—to three dimensions of space, they are chiral, but would cease to be so if transported to a space of four dimensions. So whether or not they are chiral depends on a feature of space itself: its dimensionality. This gives us a new argument, closely related to the argument from handedness, but without the latter's false assumptions.

The argument from chirality

1. There is an objective difference between a universe containing only a chiral hand and one containing only a non-chiral hand.
2. Chirality is a spatial property.
3. The spatial properties of an object are determined by either (a) relations between that object and other objects, or (b) relations between the parts of the object, or (c) a combination of (b) and some feature of absolute space.
4. In the two universes, chirality cannot be explained by (a), as there are no objects besides the hand.
5. It cannot be explained by (b), as these relations are the same in the two universes.

Therefore:

6. It must be explained by (c) (where the feature of space in question could be, for example, its dimensionality).

Therefore:

7. Handedness requires the existence of absolute space.

I have presented the argument in a way that mirrors the argument from handedness as closely as possible, but we do not really have to imagine a universe containing a single object to argue from chirality to absolute space, as we can see (at least in the two-dimensional case) that other objects are not relevant to determining the chirality of an object.

So Kant was right that a characteristic feature of the hand depends on space as a whole, but he chose the wrong feature of hands, namely their handedness, rather than their chirality. In addition, he did not identify the aspect of space on which chirality (partly) depends, namely its geometrical properties. Once again, the

Leibnizian challenge to the absolutist to show that space itself is capable of explaining anything can be met. A proper understanding of handedness and chirality must appeal to space itself.

I suggested that this feature of objects, unlike motion, had nothing to do with forces. In a sense, that is right, but the relationist may still appeal to forces in order to avoid the conclusion of the argument from chirality. Can the relationist explain the difference between three-dimensional and four-dimensional space? Not by reference to objects, if we want to allow that a four-dimensional space can be inhabited solely by three-dimensional objects. But objects, whatever their dimensions, have characteristic forces around them, and the relationist can exploit this in the current debate over chirality. We suggested in Chapter 3 that the force fields around objects can play the role absolute space does for the absolutist. Now, as the absolutist would put it, the nature of these force fields depends on the dimensionality of space. Newton's inverse square Law of Gravitation states that gravitational force exerted by an object will vary inversely with the square of distance from the centre of that object. But this is only true in a three-dimensional space. In a two-dimensional space, it would vary inversely simply with the distance, and in a four-dimensional space, it would vary with the *cube* of the distance. For the relationist, then, a possible strategy would be to say that the dimensionality of space just is the way in which gravitational force varies with distance. As before, the absolutist will insist that there is a deeper level of explanation (that it is a feature of space that explains which law of gravitation is correct); and, as before, the relationist will resist the pressure to descend to it.

Finally, we can return to the puzzle with which we began this chapter. How do we escape from the locked room? Again, we need to bring in dimensionality. Consider first a two-dimensional prison, a square (Figure 5). Our two-dimensional figure within cannot escape, it seems, as the walls around him are continuous. But if we permit him to move in three dimensions, he can easily escape with-

Fig. 5. The two-dimensional prison

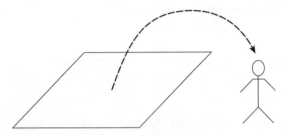

Fig. 6. Escaping from the two-dimensional prison

out breaking the boundary (Figure 6). The prison is closed in two dimensions, but open in a third.

Now consider a three-dimensional prison, a cube (Figure 7). Again, it would appear that the figure within cannot escape, surrounded as he is on all three sides. But what if there were a fourth dimension, and that the cube, although closed in three dimensions, was open in the fourth? Then our hero can escape without breaking the walls of the prison. It would appear, then, that we live in a space of only three dimensions, otherwise we would from time to time hear of mysterious breakouts from our places of confinement.

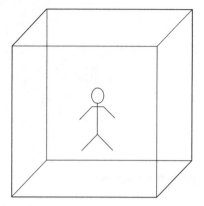

Fig. 7. The three-dimensional prison

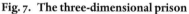

Questions

'Flatland' is a universe where space has only two dimensions. How would things look to a Flatlander? What would they make of the idea of a third dimension of space? How would they understand the idea that their space was curved?

What exactly is the difference between your right hand and your left hand?

A Möbius strip is made as follows: take a long strip of paper, give it a single twist, and join the ends together. How many surfaces does the resulting object have? Now place the shapes in Figure 5 on the Möbius strip. Without letting either leave the surface of the strip, can you make them look exactly the same?

CHAPTER 5

The Beginning and End of Time

I am Alpha and Omega, the beginning and the end, the first
and the last.

The Revelation of St John the Divine

Echoes of Creation, Portents of Armageddon

In the 1920s, the American astronomer Edwin Hubble was working
at the Mount Wilson Observatory in California. He had joined the
staff there after the end of the First World War, at the invitation of
his old university teacher at Chicago, George Ellery Hale, although
at one stage Hubble had contemplated a career in law. Mount
Wilson was a remarkable observatory in a number of ways: it had
been built on a mountain top, and so had a clearer view of the skies
than observatories situated lower down, where the atmosphere is
denser and so filters out more of the light reaching us from space. In
addition, it could also at that time boast the largest telescope in the

world: a 100-inch reflector. Hubble's first discovery was that there were galaxies beyond our own: collections of stars arranged as giant spirals slowly rotating about their centre, some of them more than 100 million light-years away. When Hubble then studied the spectra from these galaxies, he found something remarkable. (A spectrum results when light is passed through a prism, thus splitting the light into its constituent colours.) The spectrum of light from our own sun exhibits certain dark lines: these represent the wavelengths of light that are absorbed by elements in the sun's atmosphere. Hubble found that spectra of light from distant galaxies showed the same characteristic set of lines as from stars in our own galaxy, except that they were shifted towards the red end of the spectrum. This is known as the 'red shift' phenomenon. Hubble's explanation of this was that distant galaxies were receding from us. For, although the speed of light is unaffected by the speed of its source, its frequency and wavelength are not. (Light travels in waves. The *frequency* of light is the number of waves passing a particular spot per unit of time. The *wavelength is* the distance between waves. The relationship between velocity, frequency, and wavelength is given by $fl = v$.) Light from a source moving away from us will have a lower frequency than light from a source moving towards us. The lower the frequency, the larger the wavelength (given that speed is constant), and red light has a larger wavelength than blue. So light from a receding source will be shifted towards the red end of the spectrum, and light from an advancing source will be shifted towards the blue end. We have all experienced the acoustic version of this phenomenon, when the pitch of an approaching motorbike's drone suddenly drops as the motorbike passes us and speeds off into the distance. Hubble's next discovery was that, the further away the galaxy, the faster it was receding from us, since the spectra exhibited a larger red shift than those of nearer galaxies.

What are the implications of an expanding universe? If galaxies are receding from each other, then they can hardly have been doing this forever. Suppose we are watching a vastly speeded-up film of the universe running backwards. What we see are the galaxies getting closer and closer. Eventually, all matter is concentrated into a tiny space. Now we run the film forwards and we have a picture of what was memorably described by the British cosmologist Fred Hoyle as 'the Big Bang'. (Ironically, since his name for it has stuck, Hoyle himself did not believe in the Big Bang, and developed an alternative 'steady state' theory of the universe.) Thus the red shift phenomenon provided evidence (now regarded as equivocal) that the universe had a beginning. For some, this is the moment of creation.

Assuming that the universe is indeed expanding, we may wonder if this expansion will go on forever. Whether it does depends just how fast the initial expansion from the Big Bang was. If it was fast enough, then it would have overridden the gravitational forces each object exerted on others, and the expansion would carry on indefinitely. But at lower speeds, gravitation would exert its effect and cause the expansion to slow down, eventually bringing it to a complete standstill. At that point, the continued gravitational attraction between galaxies will bring them closer and closer together, resulting in a mirror-image of the Big Bang: the 'Big Crunch'. Thus, we have a picture, not only of how the universe began, but also of how it might end. This is an extraordinary advance in our understanding of the universe, but the question we need to ask is: does the beginning and end of the universe imply the beginning and end of time itself? This is a philosophical question, and it is not at all obvious that the answer must be 'yes'.

Here is a conceptual reason in favour of identifying the Big Bang with the beginning of time. Recall the sufficient reason argument introduced in Chapter 2, an argument against the existence of temporal vacua in the past. Putting it in the context of this discussion, if the Big Bang was preceded by aeons of empty time then we lack a sufficient reason for it: we would have to accept that the Big Bang was an uncaused, and so random and inexplicable, event. (The difficulty here echoes a problem St Augustine wrestled with: what was

God doing before he created the world? Augustine dismisses contemptuously the wits who reply 'He was preparing Hell for those who pry into such mysteries.') If the Big Bang marked the beginning of time itself, however, then we are not left with the embarrassment of a temporal vacuum before the world began.

Against this line of reasoning is the following cosmological speculation. The obvious similarity of the Big Bang and Big Crunch suggests that the Big Bang may itself have been the result of the collapse of an earlier universe. Indeed this cycle of Big Bangs and Big Crunches could have been repeated—and may continue to be repeated—*ad infinitum*. Any evidence for the Big Bang will not therefore tell us whether or not it was truly the first event, or merely one of a never-ending series of such events. As Stephen Hawking points out, the theories we rely on in inferring the Big Bang break down at just that point, so we cannot extrapolate back to a time before the Big Bang.

Similar considerations apply to Big Crunch: we cannot say that this would not be followed by another Big Bang, and so on forever. In any case, even if we could discern some reason why the Big Crunch had to be the final event of history, there is not the same conceptual bar to a never-ending temporal vacuum following the end of the universe as that to one preceding its beginning. We should be wary, then, of assuming the fate of our universe to be the fate of time itself.

The Limits of Reason

If anything, the weight of conceptual reasons seems, at least at first sight, to be *against* a beginning or end to time, in that the very idea of them invites paradox. Events have beginnings and endings in time, but how can time itself have a beginning or end? A beginning implies a change: some thing or state of affairs now exists that did not exist before. So a beginning to time suggests that, whereas time now exists, it did not do so before. Before what? Before time itself, presumably. But this is nonsense: there are no moments before the beginning of time. The paradox here, however, is an entirely specious one. At most it shows that we should avoid describing the beginning of time as a change. A more satisfactory expression of the idea is this: time has a beginning if there was a first moment, or, alternatively, if the past is only finite in extent. This definition will, as it turns out, have to be modified, as, for reasons we shall encounter later, a beginning to time does not entail that there was a first moment, and the past can be finite without there having been a beginning. But we shall ignore those subtleties for the time being.

A slightly more respectable reason for rejecting a beginning (and end) to time can be found, once again, in Aristotle's *Physics*. In Book VIII Aristotle writes:

[S]ince time cannot exist and is an unthinkable part from the now, and the now is a kind of middle-point, uniting as it does in itself both a beginning and an end, a beginning of future time and an end of past time, it follows that there must always be time. (Hussey 1983, 251^{b}19–22)

Representing this as a valid argument against a first moment, we have something like the following:

1. Every moment is (at some time) present.
2. The present is the boundary between past and future.
Therefore (from 2):
3. If there was a first moment of time, it could not have been present—because there can be no *past* at the beginning of time.
Therefore (from 1 and 3):
4. There was no first moment of time.

Premiss 1 seems to be uncontroversial, although in fact it makes a presupposition that we will challenge in Chapter 8. Premiss 2 however seems to beg the question. Unless we want automatically to dismiss the possibility of a first moment, we should in fairness qualify 2 as follows: *if time has no beginning* (and no end) then the present is invariably the boundary between past and future. Alternatively, we could define the present as immediately after the past, or immediately before the future, or both. So there is no very compelling argument here to suggest that time's unboundedness follows from the definition of the present moment.

In the previous section, we presented the demand for a sufficient reason as putting pressure on us to identify the beginning of the universe with the beginning of time. But it is perhaps more appropriately used *against* the notion of a beginning. For suppose time, and therefore the universe had a beginning. Then, although we cannot legitimately ask why the universe did not begin earlier than it did (for there was no earlier for it to begin in), we can still ask why it began in the *way* that it did. For there are, presumably, any number of non-contradictory accounts of how the universe might have started. Why did it start *this* way? If the universe and time began at the same point, we cannot answer this question, for any attempted answer would have to appeal to some antecedent state, and that is precisely what we have ruled out: no state precedes the beginning of time. So if the demand for a sufficient reason is legitimate, and we are correct in supposing that a sufficient reason requires that every state of the universe has an antecedent cause that explains it, then we have a purely conceptual reason for supposing that the universe, and so time, had no beginning.

We have to ask, however, whether the demand for a sufficient reason is always legitimate, and whether meeting it always means pointing to a cause. Taking the second part of the question first, we can readily admit that not all explanations are causal. We can explain some things by showing what follows logically from what, or what is required by the definition of a term, or in terms of purpose. But when it comes to explaining a state of affairs which could have been otherwise, and which is plainly the kind of thing that can be the effect of a cause, a causal explanation seems the obvious candidate. As to whether the demand for a sufficient explanation is always legitimate, the answer is that, unless the range of cases where the demand is made is restricted in some way, it cannot always be met, *whatever* view we take of time and the universe. We have just seen that a beginning of time appears to violate the principle that everything has a sufficient reason for its existence. But now let us suppose that there was no beginning, and that time and the universe have existed for all eternity. Is there anything for which an explanation could not, even in principle, be found? We can, it is true, offer an explanation of every individual state of the universe in terms of an antecedent state, but what we cannot do is offer an explanation of the *series of states as a whole*. For we can imagine another universe whose extension in time is just as great as that of our universe, and yet whose history is totally different. Why then has our universe had exactly the history it has had, and not some very different one? We cannot answer this question simply by explaining each individual state, for that presupposes a given history. It is precisely this kind of reflection, that there is no adequate explanation for the nature and existence of a universe that has no beginning in time, that is the thinking behind the first cause argument for the existence of God: since an infinite series would have no explanation for its nature or existence, the universe cannot be an infinite series of states, but must have had a first cause, a cause that does not itself require an explanation, and this is what we mean by God. Well, I do not endorse that argument, but I mention it to show what a very equivocal tool the principle of sufficient reason is when it comes to deciding whether the universe must, or cannot, have had a beginning.

Can the Past be Infinite?

We encountered Kant in the last chapter. It was then the 1760s, and Kant was pondering the significance of hands for the nature of space. Then, it seemed, he was prepared to countenance the existence of space as something independent of the world and the minds within it. But the scene has now changed. It is now 1781, and Kant has occupied the Chair of Logic and Metaphysics at Königsberg for over ten years. His lectures have become famous for their passion and eloquence. And in this year he publishes the work that was, eventually, to make his international reputation: *The Critique of Pure Reason*. Here he expounds his, by now very different, views on time and space. In the second half of the *Critique* we find the following argument:

If we assume that the world has no beginning in time, then up to every given moment an eternity has elapsed, and there has passed away an infinite series of successive states of things. Now the infinity of a series consists in the fact that it can never be completed.... It thus follows that it is impossible for an infinite world-series to have passed away, and that a beginning of the world is a necessary condition of the world's existence. (Kant 1787, A426)

Intriguingly, this argument appears side-by-side on the page with an argument for precisely the opposite conclusion, that the world had *no* beginning in time! That second argument is, in effect, the sufficient reason argument against temporal vacua we considered in Chapter 2. What is Kant up to? Can he really be promoting a self-contradiction, that the world both has, and has not, a beginning? Of course, he is doing nothing of the kind. This conjunction of arguments is the first of his four 'Antinomies of Pure Reason'. Besides this pair on the boundaries of space and time, he presents arguments for and against the propositions that substances are composed of indivisible parts, that our actions are free, and that there exists a necessary being. The conclusion of one argument of each antinomy he calls the *thesis* and the conclusion of the other the *antithesis*. The reason we get into difficulties, Kant explains, is that we take the antithesis to be the *contradictory* of the thesis: i.e. we assume that they cannot both be true *and* that they cannot both be false. But, says Kant, both thesis and antithesis rest on an assumption we can reject. There is, in other words, a third position we can take. For instance, in the First Antinomy, we assume that time is something out there in the world, a property that things really have.

So either the world has, objectively speaking, a beginning, or it has no such beginning. But we are making a mistake. Time is not something independent of us. It is a framework—what Kant calls *a form of intuition*—which we impose (unconsciously) on experience in order to make it intelligible. The same is true of space. It is not that we have a choice in the matter: we order experience both temporally and spatially quite unconsciously most of the time, and if we did not do so, it would make no sense to us. But since the temporal order and extent of things is not something outside of us, we are no longer faced with a choice between regarding the world as temporally finite and regarding it as infinite in extent: it is neither. Another way of putting the matter is that if we do, erroneously, treat time as something independent of us, we are forced into contradiction: the world is both finite and infinite. Time, then, is in the mind. That is Kant's remarkable conclusion.

Let us look again at one half of the First Antinomy: the argument for the world's having a beginning in time. We have been careful in this chapter to warn against an unthinking identification of the beginning of the universe with the beginning of time itself. But Kant's argument, if sound, applies as much to time itself as to the universe; so let us present it in those terms. Since it depends on a defining feature of the infinite, namely its uncompletability, we will call it 'the argument from uncompletability'.

The argument from uncompletability
   1. If time had no beginning, then an infinite amount of time has already passed.
   2. If an infinite amount of time has passed, then it is possible for an infinite series to be completed.
   3. It is not possible for an infinite series to be completed.
   Therefore:
   4. An infinite amount of time has not passed.
   Therefore:
   5. Time had a beginning.

We will have something to say about premiss (1) in the next section, but our current concern is with (3). It certainly seems plausible enough. Imagine any infinite task—counting to infinity, giving a complete expansion of π, etc. Could we complete any of those tasks? No. But notice that when we think of an infinite task, we naturally think of a task that has a definite beginning but no end. We define the task, in fact, by an initial position and an operation to be performed on that initial position. But series which have a beginning but no end are only one kind of infinite series. There are those that have neither a beginning nor an end, and those that have an end but no beginning. Consider the series of negative integers (. . . −4, −3, −2, −1). It has an end, a last number, namely −1, but no beginning. Wittgenstein once invited a lecture audience to imagine coming across someone apparently talking to himself. As we get nearer, we hear the words '5, 1, 4, 1, 3—finished!' We ask him what he has been doing, and he replies that he has just been reciting the complete expansion of π backwards. 'Impossible!', we exclaim, 'When did you start?' He seems puzzled, and explains that there was no point at which he *started*. If there had been such a point, he would have had to have started with a particular integer, but the expansion of π has no last integer. So he never started, but has been counting down from all eternity. However far back we go in time, he has been counting. *That* is why he has been able to complete an infinite series. Strange, most certainly, but not actually *impossible* (logically impossible, that is).

So the mathematical response to Kant's argument is that there are some infinite series that have an end but no beginning, such as the series of negative integers, and if being completable simply means having an end, then some infinite series can be completed, and so it is not, after all, a defining property of an infinite series that it cannot be completed. The premiss to reject, it seems, is (3).

However, although this is the technically correct view to take of infinite series, is it an appropriate response to this particular argument? For this argument is concerned, not with an abstract number series, but with an actual process, namely the flow of time itself. And when we are considering such processes, is it not natural to assume that they have a beginning? Now, admittedly, in trying to explain what it was for time to have a beginning, we exploited a mathematical analogy: the series of positive integers has a beginning in that it has a first member; similarly, time has a beginning if there was a first moment. So, if a number series can be one that has an end but not beginning, why not time? Well, perhaps it is because natural processes take place in time, and our reluctance to accept natural processes that have no beginning may be connected with our demand for explanation. The problem with a series of states that has no beginning is that the causal explanation of each state cannot be finite: it will go on forever. So any causal explanation of a given state in the series will necessarily be incomplete. We are, then, faced with a rather uncomfortable dilemma. If we say that time had a beginning, then we imply the existence of an uncaused event, namely the beginning of the universe (the result is the same, whether or not we identify the beginning of time with the beginning of the universe). But if instead we say that time had no beginning, then we imply that the past extends indefinitely, and there can be no finite explanation of the existence of any stage in the universe's history. Is there any way out of this dilemma?

The Great Circle

Embroidered on the chair of state of Mary Queen of Scots, in the Royal Palace of Holyrood House in Edinburgh, is the motto 'En ma fin est mon commencement': in my end is my beginning. These same words appear in T. S. Eliot's poem, 'East Coker', one of his series of poetic meditations on the nature of time, *Four Quartets*. The following lines of the poem seem to be an explicit reference to the source of those words:

> And a time for the wind to break the loosened pane
> And to shake the wainscot where the field-mouse trots
> And to shake the tattered arras woven with a silent motto.

The motto has a peculiar significance for Mary's story, for although she was to die in exile, as a traitor, and long since deposed from her throne, it was she, not Elizabeth, who was to provide Elizabeth's successor, James, King both of England and of Scotland.

'East Coker' begins with images of cycles: the cycle of the seasons, the decay of living things into the matter from which they were made, the crumbling and replacement of houses, the circular movement of a country dance. In all these, there are no ultimate beginnings nor endings, just stages in an endlessly repeating pattern of changes. Or, as we might put it, the end of one stage is merely the beginning of another. And so it is, Eliot may be saying, with human life: the end of our natural lives is but the beginning of a new existence in the life of the soul.

Well, perhaps, but theology aside, there are two ways of interpreting the image of time presented in 'East Coker'. One, the more obvious interpretation, is that history—the series of events in time - is cyclic, so that whatever is happening now will happen again at some point in the future. But time itself remains linear: each cycle, although very similar to its predecessors, will nevertheless be taking place at a later time, and the individual events, although not different in kind from earlier events, will nevertheless be numerically distinct. One and same event cannot happen twice, although one and the same *kind* of event can be instanced more than once. Each spring is a different spring. The second, perhaps more intriguing interpretation (although I do not seriously suggest that this is what Eliot had in mind) is that it is not merely history, but *time itself* that is cyclic. In other words, when we order events in time, according to what came before what, the result is a great circle (Figure 8). Notice that each event occurs precisely once, history does not repeat itself, and yet there is neither a beginning nor an end. For every time, there is both an earlier and a later time. Time on this picture is both *unbounded* (it has neither a beginning nor an end) and yet *finite*: it is not infinitely extended in either direction. So cyclic time avoids the difficulty associated with the notion of a beginning of time—that there would have to be an uncaused event—for on this picture each stage in history is preceded by an earlier stage that may be regarded as its cause. But it also avoids the difficulties associated with the idea of an infinite series—that there can never be a complete causal explanation of any given stage in history—for we need only specify a finite chain of causes for each event: ultimately, the chain of causes, if extended far enough, will come back to the event with which we began.

If cyclic time is a coherent possibility then we can agree with Kant that we are not compelled to choose between saying that the world had a beginning in time, and saying that the world's past is infinite in extent. There is a third possibility: that the past is both finite and yet has no first moment. And this very possibility allows us to avoid Kant's alternative: that time does not belong to the world in itself. However, before we happily adopt this resolution of our difficulties, we need to face the new difficulties this remarkable picture of time brings with it.

The first, not entirely articulate, expression of puzzlement is just incredulity. How can what lies in the past also lie in the *future*? How can what is happening *now* be also distantly *past*? How can it be that we are remembering the future? Perhaps the best way of dealing with this worry is to point out that, if time is a circle, it is a very large circle indeed, given that light is now reaching us from galaxies that

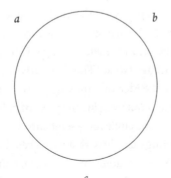

Fig. 8. Cyclic time

are 10,000 million light-years away. So there is still what one might call a *local* asymmetry between past and future: what is in the very recent past is not also in the near future; what is happening now is not even in the relatively distant past, and what we remember is not what is just about to happen.

A second, rather more precise, worry is that cyclic time may seem to be at odds with one of time's most distinctive features: its direction. For if past and future are one and the same, does that not imply that the direction from past to future and from future to past is one and the same, and that therefore there is no fundamental direction? To put it another way, the direction of time is the difference between past and future. But if time is cyclical, there is no such difference. This line of thought is, however, mistaken. The notion of a cycle is not incompatible with that of direction. Think of Eliot's country dancers linked in a circle. Each may be looking at the partner on their right rather than their left; the ring of dancers may be circling the fire clockwise rather than anti-clockwise. And, as pointed out above, there is a local asymmetry in cyclic time, in that the recent past is not also the recent future. But there is one way of expressing the direction of time that is clearly inconsistent with cyclic time, and that is the suggestion that, while the past is real, the future is not. (An issue we take up in Chapter 8.)

A third worry concerns the passage of time. Inevitably, we imagine the present moving around the circle: first *a* is present, then *b*, then *c*. Eventually the present comes back to *a*. But then does that not mean that *a* becomes present twice and so, absurdly, that one and same event occurs twice? This is clearly incoherent, and yet once we introduce the idea of a moving present into the picture of cyclic time, we cannot but imagine the present going around the circle repeatedly, and if the circle represents time itself then we have to say, thus contradicting ourselves, that each event happens both once and an infinite number of times. We are, in fact, importing *two* representations of time into the picture: the circle itself, and the motion of the present around it. But we cannot, it seems, have both. So there appears to be a tension between the idea of cyclic time on the one hand and the passage of time on the other.

The fourth and final difficulty concerns causality. We advertised it as an advantage of the cyclical time model that, for any event, there is only a finite chain of causes leading up to it, but still no uncaused event. But it is surely very odd that the chain of causes should lead us back to the very event we were trying to explain. For no event, surely, can be the cause—even the distant cause—of itself. To put the matter a little more formally: we standardly think of the causal relation has having the following logical properties: *asymmetry* (if x is a cause of y, y cannot be a cause of x), *transitivity* (if x is a cause of y and y a cause of z, then x is a cause of z) and *irreflexivity* (x cannot be a cause of itself). But in a series of events in cyclic time, causality cannot have all three properties. Suppose we assume it is transitive. Then, in Figure 8, if a is a cause of b, b a cause of c and c a cause of a, it follows both that a is a cause of a—so causation cannot be irreflexive—and that b *is* a cause of a—so causation cannot be asymmetric. Conversely, if we assume that causation is irreflexive and asymmetric, it cannot be transitive. So cyclic time conflicts with our ordinary understanding of causality.

What we have discovered in this chapter is that whatever the structure we suggest for time, whether we think of it as having a beginning and end, or as infinitely extended, or as unbounded and finite (cyclic), we run up against difficulties to do with causation. What we have found, in fact, is that we cannot hold all of the following assumptions about causation to be true:

(a) the chain of causes for any given event is finite;

(b) there are no uncaused events;

(c) causation is asymmetric, irreflexive and transitive.

Together, these rule out all the possible structures for time and history. Suppose that both time and history are infinitely extended in the past. This is ruled out by (a). Suppose, next, that time had no beginning, but that the universe did. This is ruled out by (b). (We cannot attempt to get around the problem by supposing the first event to be *self-caused*, as this is ruled out by (c).) Suppose that both time and the universe had a beginning (this may or may not be the same moment). This is also ruled out by (b). Suppose that time is unbounded and finite, but that history has both a beginning and an end. This, yet again, is ruled out by (b). Suppose, finally, that both time and the universe are unbounded and finite. This is ruled out by (c).

So our choice is now clear: either we accept that the notion of time involves inescapable paradox, and that therefore, as Kant argued, time does not belong to the world itself but is rather our projection onto the world, or we must modify one or more of our assumptions about causality to make room for a possible structure for time.

Questions

Does the evidence for the Big Bang discussed in the first section also suggest that we are in the *centre* of an expanding universe?

If no collection of things can be larger than an infinite collection, does it follow that an infinite collection cannot be added to (because it is already as large as it can be)? If so, does that mean that, since the collection of past times is always being added to, it must be only finitely large, and so have had a beginning?

If time is cyclic, what becomes of the idea of a Creator?

CHAPTER 6

The Edge of Space

I could be bounded in a nut-shell, and count myself a king of
infinite space, were it not that I have bad dreams.

William Shakespeare, *Hamlet*

Archytas at the Edge

Could space have an outermost limit, a boundary beyond which
there is nothing, not even space itself? Suppose that it does and that
you are at that boundary. Could you thrust your arm beyond it? This
simply stated but baffling conundrum was first posed, so far as we
know, by a contemporary and friend of Plato's, Archytas. Besides
being a philosopher and distinguished mathematician, Archytas was
an administrator of some significance in Tarentum (now Tarento)
in Southern Italy in the fourth century BC. One of his achievements
was the construction of a mathematical theory of music. But it is for
his puzzle concerning the edge of space that he is best remembered.

Archytas' own answer to the riddle was that it would plainly
be absurd if we could not put out our arm beyond any putative

boundary; but, if we could put it out, this would show that there was space to put it into, and so what we thought was the boundary of space was not in fact so. Now suppose we shift (in thought) to the point at which our arm reached and run the experiment again. As before, we ask whether we can extend our arm beyond *this* point, and so on. Since the argument can be run indefinitely, it follows that space has no edge, and so is infinitely extended.

Why did Archytas think it absurd to suppose that one could not extend one's arm beyond any apparent boundary? We cannot be sure but, in a later elaboration of the argument, it is suggested that any failure to extend our arm beyond a given point would have to be explained by the presence of a physical barrier, and since this would necessarily take up space, it follows that there would have to be space beyond that point. So, whether or not we can extend our arm beyond any given place, there must be space beyond—to accommodate our arm if we can, or to accommodate a physical barrier if we cannot. This is questionable, however. First, we could say that the barrier that stops our arm has no thickness, like the two-dimensional surface of a sphere. Second, even if it must have a thickness, there is no reason to suppose that it has to take up more than a finite amount of space. And a finite barrier seems enough to block the argument from proceeding any further, since we are not allowed to imagine ourselves moving through this barrier. But perhaps we can try to extend the argument in a slightly different way. Suppose, then, that there is a physical barrier surrounding the universe, preventing anything going beyond it. Is the universe able to *expand*—i.e. is the barrier itself capable of increasing in size (in terms of its surface area)? If it is, then there must be space beyond the initial limits of the boundary, allowing the universe to expand into that space. But if it is not, then there must be a space-filling barrier beyond the first barrier, preventing the latter from increasing in size. Is this barrier capable of expanding? If so, then there is space beyond it; if not, then there is yet another barrier, etc., etc.

Archytas may not have been thinking of a physical barrier at all, of course. His reasoning may have been rather simpler, along the following lines: if, *ex hypothesi,* there is nothing beyond the edge, then there is nothing to prevent me extending my arm beyond the 'edge', and so it is absurd to suppose that I cannot do so. (In saying that there is *nothing* beyond the edge, we have already ruled out the existence of a physical barrier.) Later commentators, such as Alexander of Aphrodisias, were not slow to point out the flaw in this reasoning. Although, in one sense, there is nothing preventing me putting my arm out, there is nothing *enabling* me to do so either. If there is no space beyond the edge (and it could not be called 'the edge of space' if there were), then there is nothing to put my arm into. The impossibility here is a logical, rather than a physical one. Necessarily, my arm takes up space, and so requires space to move into. What is preventing me, if we can talk of prevention in this context, is a logical, rather than a physical, barrier.

Perfectly correct though this reply to Archytas may be, it does not entirely dispose of his argument. For puzzlement remains: how *would* moving objects behave at the edge of space? In Newtonian physics, an object moving with constant velocity in a straight line will continue to move in this way unless acted upon by a force. But now, let us imagine, our constantly moving object has encountered the edge of space. What will it do? No *forces* act upon it, for the mere absence of space is incapable of exerting forces. Yet it cannot continue on its straight path: it must stop, or turn around, or . . . Well, we are not quite sure how exactly it will behave, but it will certainly not behave in accordance with Newton's First Law of Motion. So we either have to abandon the First Law in favour of something perhaps more complicated, or give up the idea that the laws of motion are the same everywhere, or refuse to countenance the idea of an edge to space. So if Archytas has not succeeded in demonstrating that the notion of an edge to space results in logical absurdity, he has, it seems, succeeded in making us uncomfortable with the idea.

Is There Space beyond the Universe?

Assume, then (for the time being, at any rate), that space has no edge, and that it is infinite (these are not equivalent ideas, as we shall see in the next section). Is it still possible for the universe to finish somewhere—i.e. is there a collection of furthest objects from us? This is, perhaps, the most natural view to take of the world, that the universe, considered as a collection of physical objects, is limited, but space not. Yet this natural picture raises interesting conceptual difficulties.

We turn again to the First Antinomy in Kant's *Critique of Pure Reason*. This, recall, consisted of two proofs, one of a thesis, and the other of its antithesis. The antithesis in the First Antinomy is that the 'world has no beginning, and no limits in space; it is infinite as regards both time and space.' Let us concentrate just on the spatial part of this proposition. Kant's proof of it is as follows:

[L]et us start by assuming the opposite, namely, that the world in space is finite and limited, and consequently exists in an empty space which is unlimited. Things will therefore not only be related *in space* but also related *to space*. Now since the world is an absolute whole beyond which there is no object of intuition, and therefore no correlate with which the world stands in relation, the relation of the world to empty space would be a relation of it to no object. But such a relation, and consequently the limitation of the world by empty space, is nothing. The world cannot, therefore, be limited in space: that is, it is infinite in respect of extension. (Kant 1787, 397–8)

Kant begins with the supposition he wishes to disprove—that the world is spatially finite. His two premises, not here defended, are, first, that space itself is infinite; and second, that spatial relations hold only between objects (as opposed to positions in absolute space). He clearly takes it as self-evident that space can have no boundaries (or at least that, considered as something outside us, it makes no sense to think of space as having boundaries) and so not in need of any defence. Given the worries raised in the last section, we may not be entirely unhappy to grant him that premiss. It is perhaps more puzzling that he offers no defence of the second premiss, which is nothing less than an assumption of relationism about space. But his earlier work on incongruent counterparts shows that he was—at least at one time—seemingly attracted to the notion of absolute space. So this shift in view requires some explanation.

We might have expected Kant at this stage to appeal to sufficient reason in defence of the relationist premiss, as he does in his proof that the world cannot have had a beginning in time. For, just as a time before the beginning of the world means that there is no reason why the world began precisely when it did, so a void space outside the world would mean that there is no reason why the world is situated where it is in space rather than somewhere else. So the idea of a bounded universe in infinite space violates the requirements of sufficient reason.

The trouble with Kant's proof is that the second, relationist, premiss rather undermines the first premiss, concerning the infinitude of space. Kant uses the conjunction of these two premisses to demonstrate that the world is spatially infinite. But he could equally well have combined relationism with his thesis—that the world is spatially finite—to demonstrate that space itself is finite. For if space is nothing more than a set of relations between objects, and there are only a finite number of objects at finite distances from each other, then it makes no sense to think of space actually extending beyond the furthest objects. Interestingly, this conception of a boundary to space does not succumb to Archytas' argument. For suppose we are next to the most distant object in the universe (i.e. distant from the centre). Can we stretch out our arm beyond it? Relationism gives us no reason to suppose we cannot do so. It does, however, block the inference from the proposition that we are able to stretch out our arm to the proposition that there is *already* space there before we actually do so. For the relationist, when we extend our arm, we are creating a new spatial relation between objects, thus extending space, which is, after all, just the collection of those relations. So if we are uncomfortable with the idea of space having an edge, if we naturally assume that space must be infinite, that is perhaps because we are, unconsciously, adopting an absolutist conception of space—space as something in its own right, and utterly independent of anything it happens to contain.

As with the temporal case, Kant sets his proof that the world cannot be spatially finite side by side with a proof that it must be spatially finite. And, again as with time, from the existence of two equally compelling arguments for two incompatible conclusions, Kant draws the moral that space is not an aspect of the world existing independently of the mind. It is important to note that Kant does not take the notions of time and space as incoherent or contradictory. If that were so, it would be hard to say how they could nevertheless play a part in making experience intelligible. Rather, contradiction only threatens once we attempt to apply these notions to the world in itself. If space and time were objective, mind-independent features of reality, then we could not give a consistent account of the spatial and temporal extent of that reality.

One cannot help but wonder how Kant squares the theory of space advanced in the *Critique* with his earlier argument concerning incongruent counterparts (see Chapter 4). One might expect that the riddle of the difference between left and right would now be something of an embarrassment to Kant. But not at all. Finding that the first edition of the *Critique* was not provoking the response he had hoped for (friends told him that people were finding it unintelligible), Kant published a short work intended as a guide to some of the main themes of the *Critique,* and expressing them rather differently and, he hoped, more accessibly. This short work appeared in 1783 under the not-so-short title *Prolegomena to Any Future Metaphysic That Will Be Able to Present Itself as a Science.* Here he explicitly refers to incongruent counterparts in mathematics and nature as supporting his view of space as mind-dependent. His 1768 essay had pointed out the inability of relationism to explain the difference between an asymmetric object and its mirror-image counterpart. Subsequently, he seems to have decided that absolutism can do no better. The argument he had in mind may be something like this: the property of space that determines the handedness of a hand cannot be possessed by a mere system of spatial relations between objects, but nor can it be possessed by a thing-in-itself (why? because such a thing is just the collection of its parts, and parts of space do not determine the handedness of a hand). So, it can only be possessed by something mind-dependent.

The Illusion of Infinity

It is time to challenge the assumptions that Kant took for granted, that space is necessarily infinite. Is the infinite extent of space really any more intelligible than the notion of space having an edge? And if space is, after all, finite, can we, to return to Archytas' riddle, have any conception of how the behaviour of objects would indicate its finitude?

As we noted above, Kant has an argument against the infinite extent of the world, and it can be applied as much to space itself as to the world. His argument is as follows:

[L]et us again assume the opposite, namely that the world is an infinite given whole of coexisting things. Now the magnitude of a quantum which is not given in intuition as within certain limits, can be thought only through the synthesis of its parts, and the totality of such a quantum only through a synthesis that is brought to completion through repeated addition of unit to unit. In order, therefore, to think, as a whole, the world which fills all spaces, the successive synthesis of the parts of an infinite world must be viewed as completed, that is, an infinite time must be viewed as having elapsed in the enumeration of all coexisting things. This, however, is impossible. (Kant 1787, 397–8)

The key idea here is an account of what it is to form an idea of a very large magnitude. We take a certain unit of length (or area, or volume) that we can readily perceive, and count the number of such units required to make up the magnitude in question. We construct, in other words, the idea of the whole magnitude through the number of its parts. Thus having an idea of the magnitude requires us to be able to complete the process of counting the units. But an infinite series, by definition, is not countable: we never reach the end of the series. It follows, given the account of how we form an idea of any magnitude, that we can form no idea of an infinite magnitude.

Kant's argument may seem to prove too much, for would it not show that we cannot form an idea of the infinite at all? That really would be an absurd result, for since there is clearly no highest number, it seems to follow that the series of numbers is infinite. To deal with this objection, it may be helpful to look at Aristotle's treatment of the infinite.

Aristotle is *a finitist:* that is, he believes that it is a mistake to suppose that the number of things existing at any one time could be infinite. Now, if space were infinite in extent, then it would contain, at any given time, an infinite number of finitely large parts, and this is not possible—or at least we have no real idea of what this would mean. We might utter the words 'Space is infinitely extended', but we would have no clear idea of what they convey. However, it does not follow from this that the notion of infinity is itself an incoherent one. There are contexts where it is perfectly acceptable to describe something as infinite in certain respects. Aristotle concedes that the number series, for example, is infinite, but describes this as a merely *potential* infinity. Numbers themselves have no existence independently of the mind. What is real is the process of counting. Numbers are infinite in this sense: that whatever number one happens to have counted to, one could always count to a higher one. So infinity is not so much a property of some set of objects—the numbers—as of the process of counting, namely its not having a natural limit. But, of course, in any actual count, one can only reach a finite number. For Aristotle, there is a fundamental distinction between the potential infinite, which is perfectly acceptable, and the *actual* infinite. The potential infinite concerns things that are actualized through some process, such as counting or division, and signifies that the process has no limit. The actual infinite concerns—or would concern, if there were any such thing—things that are already actually in existence. So we can accept Kant's conclusion concerning the finitude of space without taking the argument for it to imply that nothing is appropriately describable as infinite.

Is the notion of an infinitely extended space that of an actual infinite, as Aristotle thought? Could we not adequately capture the idea of the infinite extent of space in terms of the harmless potential infinite? We might try to do so along the following lines: however far a distance you have travelled from any given object, you could always travel a further distance. But the infinite extent of space cannot just consist in this, that the process of travelling has no natural limit, *unless* we think that space is somehow actualized—brought into existence by objects travelling. In so far as we think of space existing independently of other things, its infinite extent can only be described as an actual, and not merely potential, infinite. But now suppose that we agree with Kant that space has no existence outside the mind. Then the infinite extent of space can be thought of along the lines of Aristotle's potential infinite: that however far in thought you imagine some distant object to be, you can always imagine it further away. Next, suppose we take a relationist view, and think of space as a system of mind-independent relations between objects. Then, again, we can construe the infinite extent of space as a potential infinite: however far an object has travelled from other objects, it could always travel further. The system of spatial relations can always be extended, although at any one time it will be finite. However, for the absolutist, who sees space as independent of any mind, object, or process within it, space is, if infinite, an actual infinite. It seems, then, that one's view of what it is to say that space is infinite will depend on one's theory of space, and it is the absolutist who will find it hardest to meet Aristotle's challenge to explain how, exactly, we can form an adequate conception of space as infinite.

Some universes, though finite, may nevertheless give one the illusion of infinity. Consider a hypothetical universe proposed by Henri Poincaré (whom we first encountered in Chapter 4). This universe is bounded. But as objects move away from the centre of the universe, they shrink. Elaborating on Poincaré, let us say that the universe is perfectly spherical. If the volume of any object is at the centre of the universe is v, then their volume at a distance d from the edge of the circle is given by $v \times d/r$, where r is the radius of the universe. So, in any place equidistant between the centre and the edge, objects will be exactly half the size they were at the centre.

Let us now imagine an inhabitant of this world setting off on a journey from the centre to the edge, determined to discover how far the edge is from the centre. He has with him a yard-stick to measure out his paces. To embellish the fiction yet further, we may imagine that some omniscient being has intimated to this intrepid explorer that the radius of his universe is but 10,000 yards, and the explorer wishes to verify this piece of information. He sets out, measuring each pace. But, unknown to him, both he and the yard-stick are

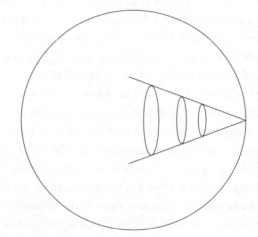

Fig. 9. Poincaré's universe

getting smaller and smaller. Each yard he measures, he is in fact traversing a diminishing distance. Of course, he eventually passes what he reckons to be the 10,000-yard mark, and seems no nearer the edge than when he started. He measures out another 10,000 yards, and still the edge is nowhere to be seen. And so it goes on. Can he ever reach the edge? No, because at the edge itself he would be, by the formula above, of zero size. He approaches that point, but never quite reaches it. Naturally, he feels cruelly misled by the omniscient being, and begins to suspect that his universe has no edge at all. But the omniscient being is no deceiver, and told him nothing but the truth.

The moral of this charming tale is that our own universe may be somewhat like this imaginary one. Perhaps space itself does have an edge, but the dimensions of objects alter as they approach the edge in such a way that they never actually reach it. We saw in Chapter 4 how objects may be affected by features of space (or, if this smacks too much of absolutism, by the features of the fields of force). Objects may alter in size quite dramatically if the curvature of space is sufficiently pronounced, as perhaps it is at the edge. So the answer to Archytas' challenge, how would objects behave at the edge of space, is: it depends what space itself is doing there.

Throughout this chapter, we have tended to identify—or at least not distinguish between—two ideas: space having no boundary or edge, and space being infinitely extended. But as we saw in the case of cyclic time, something may be both unbounded and finite. Space, too, may be unbounded and finite. It may, to use a shorter expression, be *closed*. The surface of a sphere is an example of a two-dimensional closed space: it is only finitely large, but it has no edge. The consequence is that an object setting off in a straight line in such a space would eventually return to its starting-point. So, were we to send a light beam into the furthest reaches of space (and light travels in straight lines, unless impeded or bent by other objects), would it return to its origin? Perhaps.

Questions

If space does have a boundary, what would it feel like if you tried to pass through it?

If it is possible to extend your arm beyond a given place, does it follow that there must already have been space there to receive it?

If objects did diminish in size as they approached the edge of space, could this be detected?

CHAPTER 7

Infinity and Paradox

A single room is that which has no parts and no magnitude.

Stephen Leacock, 'Boarding-House Geometry'

Zeno: How the Tortoise Beat Achilles

Achilles challenges the tortoise to a race. Being a sporting man, and recognizing his companion's inferiority in the field of velocity, Achilles allows the tortoise a head start of 100 yards. This is a more generous concession than it seems, as Achilles is not the swiftest of runners—in fact, he can only move ten times faster than the tortoise. This, however, should be more than enough to guarantee him victory. Or so, at any rate, he calculates. The race begins. By the time Achilles has reached the tortoise's starting-point, 100 yards from his own, the tortoise has moved on a tenth of that distance: 10 yards. When Achilles has covered that 10 yards, the tortoise has moved on another yard. When Achilles travels that yard, the tortoise is still ahead—by 0.1 yard. And so, to Achilles' amazement, it goes on. The tortoise always remains ahead, and although the gap between the two steadily decreases—0.1 yard, 0.01 yard, 0.001 yard—it never quite reaches zero, since one can divide any distance by 10 and go on dividing it indefinitely. The extraordinary conclusion is that if the slower runner has a head start, the faster runner will never overtake him.

This paradox is due to Zeno of Elea. He was born around 490 BC, and was a pupil of the enigmatic philosopher Parmenides, who taught that the world consisted of a single, undifferentiated object—a rather extreme form of *monism*. In his youth, Zeno is supposed to have written a book, perhaps the only book he ever wrote, presenting a variety of paradoxes associated with the idea of a plurality: that is, the idea that there is more than one object or property. Whether or not this was indeed Zeno's intention, it certainly makes sense to suppose that these paradoxes were designed to support Parmenides' monism by demonstrating the absurdities the denial of monism brings with it. Zeno and Parmenides belong to a time when philosophical ideas were only just beginning to be written down, and it is hardly surprising that no philosophical text from the period before Socrates survives. Our knowledge of them depends on fragments quoted by later writers, one of the earliest being Plato, writing in the fourth century BC, and one of the latest being Simplicius, writing in the sixth century AD. In addition, the ideas and arguments of the pre-Socratics are discussed by a number of early Greek writers, the most significant, apart from Plato, being Aristotle. In Zeno's case, we have only one passage purporting to be a direct quotation (by Simplicius); for the rest, we have to rely on reports by writers who, like Aristotle, regarded the paradoxes with suspicion.

The paradox described above, known as the *Achilles* (the introduction of a tortoise rather than an anonymous human competitor is a later embellishment), is one of the four famous paradoxes of motion attributed to Zeno. The others are the *Dichotomy* (sometimes referred to as the Stadium), the *Arrow*, and, the *Moving Rows* (also, confusingly, sometimes referred to as the Stadium). The Dichotomy is arguably just another version of the Achilles, as it seems to depend on exactly the same ideas. It goes as follows: consider an object moving from A to B. In order to reach B, it must first travel half the distance from A to B. It must then travel half the remaining distance, and then half of the distance that then remains, and so on. Clearly, this process has no end, as we can go on and on dividing the remaining distance, however small, into half. The conclusion is that in order to move from A to B, the object must travel an infinite number of half-distances, which is impossible, as nothing can complete an infinite task. Another version of the paradox, which suggests that the object cannot even get started on its journey, goes as follows: in order to reach B, the object must first travel half the distance between A and B; but in order to travel that half-distance, it must first travel half that half-distance, and so on. And so, in order to reach any point in the journey, however close to its point of departure, the object must traverse an infinite number of half-distances, which is impossible.

Infinite divisibility is also the subject of another Zenonian paradox, although this time one concerned, not with motion, but magnitude. Lacking any traditional name, it can be dubbed *Parts and Wholes*. Here is one way of presenting it. Consider a rod of some finite length. Now imagine it divided into two equal halves, and these halves halved, and so on. This process has no limit, so the rod consists of an infinite number of parts. How large, then, is each of these parts? If we say that each part has a definite, non-zero size, then, since the rod consists of an infinite number of parts of that size, the rod itself must be infinitely long. But then since all rods, of whatever size, are infinitely divisible, all rods must be of infinite length, which is plainly false. But then if each part has no size, then the rod itself can only be of zero length, since even an infinite number of parts of zero size cannot add up to something of non-zero size.

Of course, we could point out that the rod is not infinitely divisible, since eventually any process of division would have to stop at the level of the fundamental constituents of matters, if not sooner.

This is precisely what the atomists, whose pre-Socratic representatives were Democritus and Leucippus, assumed. But this move does not really defeat the paradox, for, instead of the rod itself, we could have considered the region of space the rod occupies, and we do intuitively assume that space itself is infinitely divisible.

Two Responses to Zeno: Infinitesimals and Finitism

We could respond to Zeno's paradoxes on two levels: we could view them simply as mathematical conundrums requiring a mathematical solution; or we could view them as raising deeper, philosophical or conceptual difficulties. Let us look at the first of these approaches. Zeno, we might be tempted to say, is making a simple mistake, and that mistake is most clearly seen in the Parts and Wholes paradox. Let us say that the rod (or the space it occupies) is 1 unit in length (it does not matter what unit we take). Then, if we divide it into two equal parts, each part will be exactly 1/2 unit in length. Multiplying the number of parts by the length of each part will always give us 1. So, if we say that there are an infinite number of parts, what length has each part? That is, what, when multiplied by ∞ (infinity) results in 1? The answer is $1/\infty$. Each part, in other words, is infinitely small, and the sum of an infinite number of infinitely small magnitudes is a non-zero but finite magnitude. The reason why Zeno thought that there was a genuine paradox here, according to this way of dealing with it, is simply that he (and, indeed, later writers such as Aristotle) lacked the notion of the *infinitesimal,* or infinitely small magnitude.

What of the Achilles and the Dichotomy? One of the reasons we have for thinking that, as the situation is described, Achilles can never overtake the tortoise, is that in order to overtake him Achilles would have to complete an infinite number of tasks: to run 100 yards, 10 yards, 1 yard, 1/10 yard, and so on. We also assume that nothing can complete an infinite number of tasks in a finite amount of time. But mathematically this is perfectly possible. For the sum of the never-ending series 100 + 10 + 1 + 1/10 + 1/100 . . . etc., is not infinitely large. It is, in fact 111 1/9, which gives us the precise point at which Achilles finally catches up with the tortoise. (Significantly, 111 1/9 is not finitely expressible in decimal terms: it is 111.111 recurring.) The reason that Achilles can complete an infinite number of tasks in a finite time is that each successive task takes a smaller and smaller amount of time. And so it is in the case of the Dichotomy: each half-distance we consider is half the size of the previous one, and takes half the amount of time to traverse. If we were to divide the finite distance covered by any moving object in a finite time into an infinite number of equal parts, each part would be infinitesimally small, and it would take the object an infinitesimally small amount of time to traverse it.

We have to conclude, then, that Zeno has not uncovered any mathematical impossibility in the cases we have considered. It is possible to give a coherent description of how a finitely large object can be infinitely divisible into non-zero parts, how a faster object can overtake a slower one, and indeed how objects can move any distance at all. But that is not the end of the story. For the starting-point for the three paradoxes we have considered is that space and time are infinitely divisible, and the real, philosophical problem is how this is possible. How can there exist, in any given spatial or temporal length, an infinite number of parts? The difficulty we face here is exactly the same as the one we encountered in the last chapter when we were contemplating the idea of an infinitely extended space, and the difficulty is a conceptual one. It is one thing to give a mathematical description of an infinite series, quite another to conceive of a physical realization of that description. On the other hand, what would it be to *deny* that time and space were infinitely divisible? This would imply the existence of spatial and temporal minima, that is magnitudes that could not be divided any further, and this again invites paradox (to be further examined in the final section).

However, if Aristotle is right, we really can have our cake and eat it. We can say that space and time are indeed infinitely divisible without this implying the existence of an infinite number of parts. We can invoke Aristotle's distinction between the actual and potential infinite introduced in the previous chapter. Aristotle denies that anything exhibits an actual infinity—whether this be numbers of objects, or parts of objects or of space and time. But things may be *potentially* infinite, where this is construed as the absence of a limit to some process. So let us take a given length. We can divide it into two halves, divide those halves, and so on. The process of dividing has no natural limit. (Aristotle assumed that this was true of physical objects, thus rejecting atomism, but we can restrict the assertion to lengths, areas, or volumes of space.) So that is what it is to say that a length is infinitely divisible. It does not *already contain,* waiting to be discovered, an infinite number of parts; such parts only come into existence once they are defined by an act of division. But that process does not have a pre-fixed limit imposed on it.

Aristotle's theory of the potential infinite neatly disarms the three paradoxes. The Parts and Wholes paradox depends on the idea of an actual infinite number of parts. Aristotle simply denies that there is such a collection of parts. No parts exist until one begins to divide the rod, so there is no need to address the question of how large each part is. The answer to the Dichotomy and the Achilles is similar: although we may in thought endlessly divide the motion of a body into smaller and smaller parts, these parts have no real existence unless they are marked out in some way. But the motion of bodies does not consist in a number of discrete steps: it is smooth. So in moving from A to B, an object does not need to traverse an infinite number of half distances. What, however, if we asked Achilles to mark the various stages of his journey, perhaps by sneezing when he reached the 100-yards point, then again at the 110-yards point, again at 111 yards, and so on. Then the parts of his journey would have more than a notional existence: they would, because marked out in this, admittedly somewhat eccentric, way, really exist. Would Aristotle not have to concede that in *this* case Achilles could not overtake the tortoise, for he would have had to sneeze an infinite number of times before he did so?

Thomson's Lamp

The case just imagined might make us wonder whether we cannot describe a situation, without falling foul of contradiction, in which an actual infinity of states is realized. The sneezing Achilles may seem too absurd to warrant serious consideration, but there is another possibility, first described in 1954 by James Thomson, later Professor of Philosophy at the Massachusetts Institute of Technology. Consider a lamp whose switch is controlled by a sophisticated timer. At the start of the experiment the lamp is switched on, and remains on for exactly one minute. At the end of that minute, the timer switches the lamp off. It remains off for half a minute, and then is switched on for 15 seconds, then off for 7.5 seconds, then on for 3.75 seconds, and so on. Each period in which the lamp is on (off) for n seconds is followed by a period in which the lamp is off (on) for $n/2$ seconds. Now consider these two questions concerning the state of the lamp at the end of 2 minutes: (i) how many times has the lamp been switched on and off? (ii) is the lamp now on or off?

The answer to the first question is rather startling. If we represent the successive states of the lamp by the series of increasingly short periods in which it is on or off, we obtain something that has no last member: 60, 30, 15, 7.5, 3.75, 1.875 . . . The series, in other words, is infinite. So, at the end of two minutes, the lamp has been switched on (and off) an *infinite number of times*. Now, there is nothing mathematically incoherent in the description of the experiment, for the sum of the series of periods is not infinite: it approaches, without quite reaching, 120 seconds. But is there nevertheless an absurdity lurking somewhere in the story we have told?

Consider the second question: is the lamp on or off at the end of the 2-minute period? Here it looks as if we are faced with a contradiction, or at least serious anomaly. The lamp, it seems, cannot be on, for every period in which the lamp was on was followed by a period in which the lamp was off. So being on cannot be the final state of the lamp. But, equally, it cannot be off, for every period in which the lamp was off was followed by a period in which the lamp was on. So being off cannot have been the final state of the lamp. Our reasoning has led us to the, surely unacceptable, conclusion that the lamp is neither on nor off. Here, according to Thomson, is where the absurdity of the case really lies. So even its inventor does not think that Thomson's lamp, as it is known, is a genuine possibility.

We do not have to agree with this conclusion, however. We noted above that the sum of the series 60, 30, 15, 7.5, etc. does not quite add up to 120. In other words, the series approaches the 2-minute mark without quite reaching it. Another way of putting this is that the period which begins at the 2-minute mark is not itself a member of the series of periods controlled by the timer. Nothing in our description of the situation then enables us to deduce what the lamp is doing at the 2-minute point, so we are not forced to say that the lamp is neither on nor off. The right answer to (ii), surely, is simply 'We cannot predict.'

No doubt someone will point out, rather tiresomely, that the set-up is not a genuine physical possibility. You cannot switch a lamp on and off that number of times without blowing a fuse or breaking the bulb. In any case, at some point the switching on and off will be so rapid that the electricity would have no time to flow through the circuit, and the lamp will remain effectively off. This really is of no interest at all. The lamp itself is just there for dramatic effect, as it were. We could have concentrated solely on the state of the switch. A slightly more sophisticated objection is that, if we imagine a mechanical lever switch moving back and forth, we would soon reach a point where the lever was moving at the speed of light, and nothing can surpass the speed of light. Well, maybe it is not that kind of switch. In any case, even if physics forbids the experiment reaching its conclusion, that does not make the story a contradictory or incoherent one, for we can always locate the experiment in a world in which the laws of physics are very different. What we are after is some indication that there is a logical impossibility in the story, and so far we have failed to find one. It could be objected that we do not really understand the situation. But this does not seem to be a particularly good strategy. We have, after all, described it in some detail. What part of the description is it that we supposedly do not understand?

But if there really is no contradiction in Thomson's lamp, then have we not found a counterexample to Aristotle's thesis that there cannot be an actual infinite existing in nature? Certainly the potential infinite is not sufficient to describe what is happening, for the various parts of the 2-minute period are clearly marked by the state of the lamp. Thomson's lamp, it seems, is an example of the actual infinite: an infinite series that is genuinely completed. Is Aristotle then wrong? Not necessarily. I suspect that Aristotle would say that, as he has defined it, the actual infinite is not exemplified by Thomson's lamp. For the actual infinite is something that exists *all at once*, at one and the same time. Thus a spatially infinite universe would be an actual infinite. But the series of states of the lamp is extended over time. At any one time we only have a single state of the lamp, not an infinite number of states.

This answer is a rather disappointing one. We have the distinct feeling of an issue being dodged rather than tackled. After all, have the states of the lamp not been truly actualized? Why should it matter that the infinite series of states is extended over time rather than space? (The question is not entirely a rhetorical one. A substantial answer would have to do with the peculiar status of the present: what really exists exists now because both past and future are unreal. We examine this issue in the next chapter.)

There is a more interesting, although perhaps some will find it an equally frustrating, response to Thomson's lamp, which goes as follows. There is a kind of necessity that is neither logical nor physical, namely *metaphysical* necessity. When we say that something is metaphysically necessary, we do not imply that in denying it, one would commit oneself to a logical contradiction. Nor are we saying that it is required by the laws of nature. But we are saying that, in some fundamental sense, matters could not be otherwise. An example often given of this metaphysical necessity concerns the identities of things. Suppose we discover that A and B, whom we had previously considered distinct, were actually one and the same object. For example, perhaps we have just discovered that Robert Burton, the author of the seventeenth-century bestseller *The Anatomy of Melancholy* (still, incidentally, in print), was one and the same man as the person who styled himself 'Democritus Junior'. It is, according to some philosophers, metaphysically necessary that 'Robert Burton is Democritus Junior', although we can discern no logical contradiction in the sentence 'Robert Burton is *not* Democritus Junior', and we do not suppose that the laws of physics prevent these two from being distinct.

Granting for the sake of argument, then, this notion of metaphysical necessity, it would be perfectly in order to say that Thomson's lamp, although it can coherently be described, represents a metaphysical impossibility. Indeed, this is one interpretation of Aristotle's finitism: that the actual infinite is metaphysically impossible, although mathematically and logically possible. This is a more interesting response than the one which points out merely that Thomson's lamp at least obeys the letter of Aristotle's finitism. But it remains a limited response unless some account can be given of how we can recognize metaphysical necessity or impossibility when we encounter it. For without such an account, appeal to such a form of necessity will seem a rather cheap way of winning any argument.

There is, in any case, another problem for Aristotle. He says that a length is infinitely divisible just in the sense that there is no limit to the process of dividing. This divisibility has nothing to do with the pre- existence of an infinite number of parts, waiting to be divided. Now we assume that, when he says that there is no limit to the process of dividing, Aristotle means not just that there is no limit *in thought* to the process of dividing, but that in addition there is no *natural* limit to the process. Nothing in the length itself (for Aristotle, the physical object; for us, the volume of space) prevents the process going on indefinitely. But then this must tell one something about the length itself, and what it tells one is not merely negative, i.e. that it is not composed of atoms. For one is entitled to ask what it is about the length that enables one to go on dividing, what grounds the possibility of indefinite division. Suppose we take an absolutist view of space—that is, we regard space as a thing in its own right (see Chapters 3 and 4—and that, moreover (perhaps because we see it as a consequence of absolutism), we think of spatial points as something real. Then, if space is infinitely divisible, there must be an actual infinite number of points in any length, whether or not we choose to mark them out by a process of division. Similar considerations hold if instead of absolute space we substitute fields of force, for these fields are divisible into different parts. Without the existence of such an actual infinite (or points of space, or parts of the field), is Aristotle entitled to hold that space is infinitely divisible?

At this point, we must leave the matter, to pursue other conundrums of divisibility.

A Puzzle about Transition

A train is waiting at the platform. For a while, all is still. Then the stationmaster waves his flag, and, very slowly, the train begins to move. A familiar enough scene. But below the surface lurks a subtle mystery. It is natural to suppose, before we start thinking about it, that there is a last moment of rest before the train starts to move, and a first moment of motion. But, if time is infinitely divisible, then there cannot *both* be a last moment of rest *and* a first moment of motion. The interval between any two moments, however close together, can always be further divided. One way of expressing this is in terms of what is called the *density* of time: between any two moments there is always a third. Take, then, the putative last moment of rest and first moment of motion (Figure 10). Time being dense, there is a third moment between them. So is this third moment a moment of rest or one of motion? If of rest, then the moment we previously took to be the last moment of rest could not have been so, for there was a moment of rest following it. If of motion, then, by parallel reasoning, the moment we previously took to be the first moment of motion could not have been so, for there was a moment of motion preceding it.

Assuming we can rule out the possibility that this intervening moment is neither one of rest nor of motion, we have the following result: if there was a last moment of rest, then there was no first moment of motion; and if there was a first moment of motion, then there was no last moment of rest. We have to choose one or the other: a last moment of rest or a first moment of motion. But the choice appears to be completely arbitrary.

(Perhaps we should not overlook a third possibility that offers to save honour on both sides, namely that there is *neither* a last moment of rest *nor* a first of motion. But then, is choosing this option any less arbitrary than choosing one of the first two?)

There ought, we intuitively think, to be some systematic, non-arbitrary way of deciding. But what is it? Here is one proposal: it is misleading to talk of a moment of motion in the first place, for nothing can move at a moment, if we assume a moment to have no duration. It is strictly correct to speak only of an object's occupying a certain position at a moment. (See Chapter 9 for discussion of this

'Last moment of rest'   'First moment of motion'

Fig. 10. The transition from rest to motion

controversial point.) So, instead of talking of a first moment of motion, it would be better to talk of a first moment at which the object occupies the first position of displacement from its resting position. But there is no such position—at least, not if space too is dense, and it would be odd to assume time to be dense without assuming the same for space. So, we assume that between any two points of space there is always a third. Consider, then, the location of the very front of the stationary train, which we can, simplifying somewhat, consider as a flat, two-dimensional area. As the train moves away, the front comes to occupy a different area. But between this and the area the front occupied when the train was at rest, there is always a third area (Figure 11).

So whichever position we might choose to designate 'the first position of displacement from the resting position', it will turn out not to be so, for there is always another position nearer to the resting position. It makes sense, then, to talk of a last moment of rest, for here we are talking of a definite spatial location for any part of the train, such as the front. But there is no first moment of motion, for there is no unique initial position of displacement.

Two worries arise over this solution. One is that it is, after all, acceptable to talk of motion at a moment, and we should be suspicious of a view that rules out such talk as illegitimate. Perhaps,

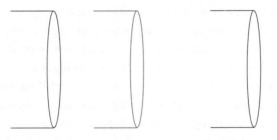

Position of the
front of train
when at rest

First position of
displacement from
resting position?

Fig. 11. The moving train

however, the matter is not so important. After all, even if we do allow talk of motion at a moment, it is helpful to think also in terms of the successive positions an object occupies at different moments. And the above solution gives us a reason for saying that, although there may be moments at which the object is moving, there is no moment that could legitimately be described as the first moment of motion.

The second worry is rather more significant. Motion is just one form of change. There are many others: change in shape, size, temperature, state, hue, brightness, pitch, volume, chemical composition, etc. The puzzle about transition from rest to motion described at the beginning of this section arises also in the context of other changes. Consider, for example, a kettle that has just been taken off the stove. Is there a last moment of the water's being at 100 °C and a first moment of being cooler? Now the solution we have just considered to this kind of puzzle, as it arises in the context of motion, depends on the fact of motion's being dense, or, to use a more familiar term, *continuous*: i.e. the journey taken by the object can be indefinitely divided. (There is a technical distinction between density and continuity, but we can ignore it here.) It seems likely, then, that we can apply the solution to other cases of change, but only where such change is similarly continuous. So, of the water, we can say that there is a last moment of its being at 100 °C, but no first moment of its being cooler if there is no unique next temperature below 100 °C. If, that is, there is always an intermediate temperature between any two temperatures we care to choose. Similarly, we can say that there is a last moment of a sound's being at a certain pitch, but no first moment of its ceasing to be at that pitch if the dimension of pitch is continuous. But is all change continuous? Perhaps some changes, such as changes in size or shape, are *discrete*: that is, for any given state along the dimension of change, there is a unique next position. As we might put it, some changes proceed in little jumps, rather than smoothly. The passage from existence to non-existence is, according to some people, one such discrete change, in that there is no intermediate state between existence and non-existence. You are either alive or not. But then we have no reason to prefer a last moment of existence over a first moment of non-existence, or vice versa.

There are three possible responses to this latest difficulty: (i) if space and time are dense / continuous, then all change is; (ii) if some changes are discrete, all changes are, and so too are time and space— a possibility we shall consider in the final section, (iii) although some changes appear to be discrete, they actually involve states whose obtaining or not is a vague matter. Thus, between definitely being alive and definitely being dead, there are intermediate states where it is just *indeterminate* (that is, there is no fact of the matter, not just that we cannot discern what the facts are) whether the organism in question is alive or not, just as it is indeterminate, in a colour spectrum, where yellow ends and orange begins.

Democritus' Cone

Our final paradox of divisibility is the cone paradox, first discovered it seems by Democritus (born around 460 BC), known as the 'laughing philosopher', perhaps because he advocated the pursuit of happiness as a morally legitimate goal in life. Not nearly so well-known as Zeno's paradoxes, it is certainly as subtle and interesting and, like Zeno's Dichotomy and Achilles, admits of a mathematical solution which yet does not diminish its philosophical interest and significance.

Picture, then, a cone, with smoothly sloping sides. Now imagine a horizontal cut, dividing the cone into two halves (Figure 12). Consider the two surfaces thus exposed, *a* and *b*. Are the areas of these surfaces equal or unequal? If equal, then the cone is not, after all, a cone, but a cylinder. For the object may be considered as a stack of surfaces, and if neighbouring surfaces are equal in area, then the sides cannot slope (Figure 13).

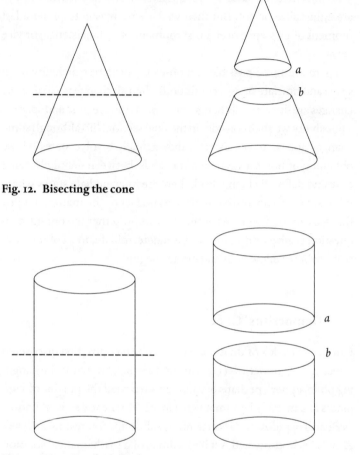

Fig. 12. Bisecting the cone

Fig. 13. Bisecting the cylinder

But, on the other hand, if *a* and *b* do differ, they differ by a definite amount, and the cone does not, after all, have *smoothly* sloping sides, but stepped ones. For, as before, the cone can be considered as a stack of surfaces, and here the neighbouring surfaces will differ by a non-zero amount (Figure 14). So, if there are cones, they must be stepped, and so consist of discrete units. Of course, this conclusion was entirely congenial to atomists (of whom Democritus was one)

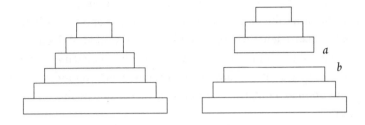

Fig. 14. The stepped cone

for they believed that matter was indeed composed of discrete and indivisible units, namely atoms (from ατομοσ, ancient Greek for 'indivisible'). But we do not have to confine this experiment to concrete, physical objects. Let us instead consider a volume of space that is cone-shaped. If the reasoning above is correct, then space itself should consist of indivisible units, which we could dub 'space atoms'. This really would be a striking result. So Democritus' cone belongs in the same family as the Zenonian paradoxes we have discussed: they all show that the assumption of infinite divisibility leads to unacceptable conclusions.

Someone, perhaps inspired by Zeno's Parts and Wholes paradox, might question the suggestion that the cone is, in effect, a pile of disk- shaped surfaces. 'For'. they will object, 'if each surface is of zero thickness, how can a pile of such surfaces add up to something that has any height?' It is not clear that Democritus' puzzle need rest on the idea of a cone being composed of surfaces, however. The point is that, if we do indeed have a cone here, then we should expect any two measurements of the horizontal area of the cone, however close those measurements are, to differ. It might be illuminating to consider similarities between the cone and the transition paradox discussed in the previous section. We suggested that, time being continuous, if there is a last moment of rest for the train, then there is no first moment of motion. So, were we to divide the period covering the train's transition from a state of rest to a state of motion into two periods, one being the period of rest and the other the period of motion, we would find that, whereas the first had a definite boundary, the second would not (Figure 15). The period of motion does not have a definite boundary in that it does not begin with a unique first moment: whichever moment we alight upon as a putative first moment we can always find an earlier one.

Now consider the cone again. We naturally assumed that, when the cone was split in half, there were two definite surfaces exposed. But were we entitled to assume that? Consider the undivided cone, but with a horizontal area identified (Figure 16). Call this plane p. Now is there a plane immediately above p? If space is continuous, no. For whatever plane we take, say p', there will always be a third plane between p and p'. Now let us split the cone, so that the lower part has p as its exposed surface. Does the upper part have a definite surface? No, for the reasons just given. For it to have a definite surface, there would have to have been, in the undivided cone, a plane immediately above p, and we saw that there was not.

Fig. 15. There is no first moment of motion

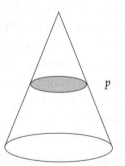

Fig. 16. The undivided cone

Now this may show that it is possible to answer Democritus' question about the cone while still holding on to the idea of space being infinitely divisible. But it still makes us uneasy about the idea of physical objects behaving in this way. For Aristotle, and other opponents of atomism, matter was continuous. So Democritus' cone is a problem for them, not just as a mathematical conundrum, or one about regions of space, but as one about physical objects. Imagine a cone being cast so that the lower half was made of lead and the upper part of gold. Aristotle would have to say that, once the halves were separated, one of them, at least, could not have a definite boundary. And this is distinctly odd. What would it *look* like? Would it look any different from the part that does have a definite boundary? In what way? What Democritus' cone teaches us, at the very least, is that Aristotle's world of continuous matter is a very strange one.

Atoms of Space and Time

There is one very simple solution to all the paradoxes of this chapter, although it is doubtful that many people would be prepared to accept it, and that is to deny that space and time are, after all, infinitely divisible. The solution is this: the process of division does have a limit, so that there is a minimum length and a minimum duration, not further divisible into smaller magnitudes. Further, between any two points of space or time there is *not* invariably a third: it depends how close the points are. Every spatial point is surrounded by a determinate number of other points, and every moment has a unique predecessor and a unique successor. Time and space are, in short, discrete.

Apart from the unfamiliarity of this picture, are there any serious conceptual problems with it? No obvious contradiction, it seems, but we can derive two interesting consequences from it. Imagine four contiguous points in space, and two objects, A and B, each only one unit in length (where one unit is the measure of a single spatial point). A and B are moving past these points, and A is moving twice as fast as B (Figure 17). Now, where will B be when A has reached the point 2? The natural answer is *halfway* between point 1 and point 2. But, if space is discrete, as we are assuming, halfway between 1 and 2 is no location at all! Where, then, is B? The answer must be: still at 1. And it remains at 1, even when B has reached 3. Only as A reaches 4 does B move to 2. The first consequence of discrete space and time, then, is that change itself would go in little jumps, that is, it would proceed from one state to the next without occupying any intermediate state, for there would be no intermediate state. Motion, for example, would be like a series of successive stills in a movie film. The moving object first occupies *this* position at *this* time, and then *that* position at *that* time. We want to ask 'But how did it get from there to here without moving through the space in between?' The answer is that there is no space in between two contiguous points in discrete space.

Here is another puzzle. Suppose *d is* the minimum length—the length of a single spatial atom. It is not possible to divide *d* and obtain a possible spatial length. Space being composed of atoms, the dimensions of any object will be expressible in integers, namely multiples of *d*. Now consider a right-angled triangle, two of whose sides have length 5*d* (Figure 18). The question is: what length is the hypotenuse? By Pythagoras' theorem, this is given by the square

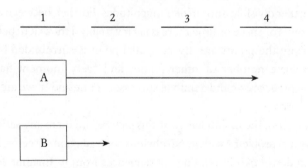

Fig. 17. Discrete motion

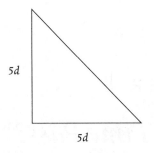

Fig. 18. A problem for Pythagoras

root of ($(5d)^2 + (5d)^2$), i.e. $\sqrt{50}$. But $\sqrt{50}$ is not a whole number, and so not permissible in discrete space. Does this show that the notion of discrete space contradicts fundamental mathematical truths, and so must be impossible? Not at all. Pythagoras' theorem is true in Euclidean geometry, but as we saw in Chapter 4, we are not obliged to take the Euclidean system as describing space as it truly is. All we can say, then, is that if space is discrete, it is not Euclidean.

Questions

Why do we find it natural to suppose that a period of time, or a region of space, can be divided up indefinitely?

A space-ship, initially stationary, begins to move at a certain moment. After 1 minute, it doubles in speed. After 30 seconds, its speed doubles again, and again after 15 seconds, 7.5 seconds, etc. Suppose its speed during the first minute was 100 miles per hour. How fast is it going two minutes after it started to move?

If between any two points in space there is always a third point, can anything touch anything else?

CHAPTER 8

Does Time Pass?

You did not come,
And marching Time drew on, and wore me numb.

Thomas Hardy, 'A Broken Appointment'

The Mystery of Passage

If one had to pick out the two most characteristic features of time, features which clearly mark its distinctness from space, one would probably cite, first, the fact that time passes, and secondly, the fact that the present is in some sense unique. These two features are closely related. For one indication of time's passing, indeed what perhaps it means to say that time passes, is that what time it is *now* is constantly changing. Watching the second hand flying around the clock face, one is acutely aware of successive times becoming present. No sooner has one registered the fact that it is 3 o'clock and time to run down to school to pick up the children, pausing perhaps to gulp down a quick cup of tea, than it seems to have come around to quarter past already and parents will already be jostling for position around the school gates, etc., etc. And so, with unrelenting speed, life's scenes fly by, and the grave yawns before us.

Well, let us not get too carried away. Perhaps time will allow us a moment of calm reflection to work out what the two features mentioned above really amount to. Since we have expressed the first fact in terms of the second, let us start by asking what it is to say that the present is unique. The contrast with space is helpful here. The spatial equivalent of 'present' or 'now' is 'here'. Is 'here' unique? Well, in one sense, it is, because it is the place where I am, and I am only in one place. So 'here' does succeed in picking out one place. But there is nothing really very special about the place where I am. There are lots of other places I could have been, and indeed lots of other places where people actually are and which they would refer to by 'here'. My place is just one among many. And suppose that there were no people around to use terms like 'here', indeed no sentient creatures at all. Would there be any place that was *here*? What I mean by this question is not, would the place at which I am at the moment exist in my absence, but rather, is the *hereness* of a place independent of anyone's actual location? The answer most people would give is no. *Hereness* is not exactly a property of places. A place is correctly describable as 'here' if it is the place where the person using the term 'here' is standing, or sitting, or whatever. So 'here' is tied to a person, and different persons have different 'heres'. No *here* is more philosophically significant, or more real, than any other. But with time we suppose it to be otherwise. That is, we naturally think that, even in the absence of any sentient creatures, some time would be *now*. Of course, nobody would be around to describe a time as 'now', but what makes a time now, or the present time, has nothing to do with whether someone is in it. *Nowness* is something that times have, independently of people, or thoughts or language. And there is only one now. (Why? We shall look at this question in the final section.) That is, reality is divided up into past, present, and future, the present being the boundary between past and future. And although different times successively become present, each time, as it becomes present, acquires a unique status: it is more significant (and perhaps more *real*) than any other moment. We do not say 'Well, this time is present for us, but for lots of other people, different times are present.' We can express the contrast between time and space in this way: what place is here is simply a matter of perspective: in describing a place as 'here' one is describing the world from one's own point of view; there are other, equally legitimate, perspectives from which this place is not 'here' but 'there'. But what time is now is *not* a matter of perspective. In describing a time as now one is not merely describing the world from one's own point of view, but describing the world as it *is*. There are no other, equally legitimate, perspectives according to which this moment is not the present moment.

That may help to explain how we think of the present, but there are some worries. Consider the words 'Only one time *is* present'. How should we read the 'is'? Here are the options: (i) as a present-tensed 'is', meaning 'is *now*', as in 'Your tea *is* on the table; (ii) as an eternal 'is', not implying that the state of affairs is happening now and only now, but that it endures throughout time, as in 'Space *is* infinite in extent'; (iii) as a timeless 'is', implying nothing about position in time, as in 'The square root of 9 *is* 3.' Now if the 'is' in 'Only one time is present' is to be read as (i), then the statement is equivalent to 'Only one time is *now* present', which, though true, does not say anything particularly significant, and has a precise spatial parallel in 'Only one place is here *here*'. If the 'is' is read as in (ii), i.e. as eternal, then 'Only one time is present' is false, for it is not true that one and only one time *remains* present for all of eternity. It is similarly false if we interpret the 'is' as timeless, for which time is present is something that changes. So it is something of a problem to explain how 'Only one time is present' can say something both true and substantial, a problem we shall return to in the last section.

Supposing we can overcome this worry we can define the passage of time in terms of the behaviour of the present: the passage of time is, as one might put it, the movement of the present through successive times and events. This characterization lends itself to metaphor. Here is one particular vivid metaphor, due to C. D. Broad:

[W]e imagine the characteristic of presentness as moving, somewhat like the spot of light from a policeman's bull's eye traversing the fronts of the houses in a street. What is illuminated is present, what has been illuminated is the past, and what has not yet been illuminated is the future. (Broad 1923, 59)

This is certainly a striking image. But is it a satisfactory one, or is it actually rather misleading? Of course, there will be limitations to any metaphor, however apt. But reflecting on the shortcomings of this one may help us to describe the passage of time a little more precisely. Some may say that we can never really go beyond the metaphor, that any attempt at a literal description of the passage of time will result in falsehood, or worse. But let us at least start by being a little more optimistic, and assume that we have some grasp of the passage of time that does not just rely on pictures.

The first shortcoming we might note is that the metaphor represents time in terms of a temporal process: the movement of the beam of light. Time (or at least an aspect of it) is being represented by . . . time. So the metaphor only works, one suspects, because the passage of time has already been built into it in a non-metaphorical way. We need to understand that the present moment is being represented by the house that *is now* being illuminated, and not one that *has been* or *will be* illuminated. A second, related problem is that processes, by definition, take time, and this means that they can speed up (take less time) or slow down (take more time). But, surely, the passage of time is not something that can speed up or slow down. For we measure the rate of change in some dimension in terms of how much time it is taking. Now we could say that the moving present took 5 minutes to move from 4 o'clock to 5 minutes past. But it could hardly have taken more or less time to do so, for 5 minutes is itself marked out by the moving now. The rate of the passage of time, then, is unvarying. But then why call it a rate at all?

Is there something perhaps deeply wrong in thinking of the passage of time as some kind of movement in the first place?

Note, thirdly, that the future in Broad's picture is represented by houses that *are already there,* waiting to be illuminated. Is this how we see the future? Do we imagine future events somehow already in existence, waiting their turn to arrive at the present? Perhaps we do, but it is not at all clear that this is how we must conceive of them. Future events, arguably, have no existence at all: they are not there in any sense. They only come into existence when they are present. Someone taking this view will think that there are *too many houses* in Broad's model.

Yet another problem, perhaps related to the third, is that we can imagine the beam moving in either direction. It is, shall we say, moving from left to right. But it could equally be moving from right to left. The order of the houses is unaffected by the direction of the movement of the beam. So some houses stand to the right of others whether or not the beam illuminates them first. But this is not how we think of time. Some events are earlier than others because they become present first. It makes no sense at all to say that the battle of Austerlitz would have been earlier than the battle of Borodino even if Borodino had become present before Austerlitz had done so. The flow of time, in the metaphor, seems to have become detached from temporal order, and the reason this is so is that the metaphor represents time *twice over.* On the one hand, we have the row of houses, representing events standing in a temporal order. But on the other, we have the moving light, representing the changing present. So, we want to say, where is time here? Is it the order of the houses, or is it the moving beam? Surely, it cannot be both.

Let us try to amend Broad's metaphor to deal with the third problem. Suppose we now say that there are *no* houses to the right of the beam (supposing the beam to be moving from left to right). They spring into existence at precisely the point at which the beam reaches that area. This represents the unreal future, and events coming into existence at precisely the point at which they become present. Becoming present just *is* springing into existence. Does this help with the other worries? Well, now the row of houses is not independent of the movement of the beam, in that the order they stand in is determined by the beam. If the beam is moving from right to left, then those houses that are to the right of others will also be the *later* houses. The first worry, however, remains. We still have to understand that the present time is represented by the house that is *now* being illuminated. But since the passage of time is a dynamic process, perhaps it is not surprising if it can only be modelled by a dynamic metaphor.

Time's passage, arguably, is what makes time *time:* it is not something that time could have lacked. But if this is right, then our ordinary notion of time is heading for trouble.

McTaggart's Proof of the Unreality of Time

In 1891, a young man of 25, by the name of John McTaggart Ellis McTaggart, was elected to a Fellowship at Trinity College, Cambridge. Something of an oddity, he had been infamous at school, both for his atheism and his anti-monarchism. These views seem to have taken root early in life. At the age of 6, he is said to have announced to an uncle that he did not believe in an after-life. He had a peculiar, side-ways walk, attributed variously to agoraphobia, repeated attacks by other boys when he was a child, and a slight curvature of the spine. The other children in the village where he grew up found him a figure either of horror or fun, with his odd walk and tendency to talk to himself, and called him simply 'the loony'. But at Cambridge he had a profound influence, while it lasted, on his younger, and now more famous, contemporaries Bertrand Russell (1872–1970) and G. E. Moore (1873–1958). McTaggart would invite any undergraduates or young dons whom he thought might be intellectually promising to breakfast, where he would regale them with his views. McTaggart's breakfasts, as Russell was later to recall, were famous for their lack of food, and it was prudent to go along armed with something to eat.

McTaggart's first meeting with Moore occurred in 1893, both having been invited to tea by Russell. During the course of the conversation, McTaggart produced his (already by that time well-known) opinion that time was unreal. Presumably, McTaggart would have produced an argument for what Moore later described as 'a perfectly monstrous proposition', yet it was not until 1908 that McTaggart's proof of the unreality of time appeared in print. Although this is the argument for which he is now famous, it is likely that he would have had more than one reason for thinking of time as unreal, but some of these may have been inextricably linked with a particular, and controversial, system of thought, influenced by the German philosopher G. W. F. Hegel (1770–1831). What is interesting about the 1908 argument, and perhaps in part why it has been so much discussed by philosophers, is that it assumes relatively little in the way of a background metaphysics.

McTaggart begins by distinguishing between two ways in which we order events in time. The first, which he called the *A-series* (Figure 19), is the series of events ordered as running from the remote past, through the near past, through the present, through the near future and to the distant future. An event's A-series position is constantly changing: it is first future, then present, then past. Once it ceases to be present, it continues indefinitely (if time has no end) to recede further and further into the past.

The second way in which we order events he called the *B-series* (Figure 20), and this orders events by means of two relations:

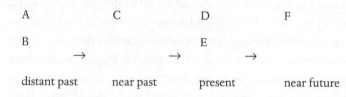

Fig. 19. The A-series

A   is earlier than   C   is earlier than   D   is earlier than   F

is simultaneous with   is simultaneous with

B   E

Fig. 20. The B-series

precedence and simultaneity. The B-series locations of events do not change: if at one time it is the case that A is earlier than C, then it is the case at all times that A is earlier than C.

We can now distinguish between two kinds of temporal expression: A-series expressions and B-series expressions. A-series expressions help to locate events in the A-series. Examples of this kind of expression are: 'The visit to the museum is taking place *tomorrow*', 'She is arriving at the station *right now*', 'We last met *ten years ago*.' B-series expressions help us to locate events in the B-series. Examples of this kind of expression are: 'The total solar eclipse takes place on *11 August 1999*', 'The passage of protons through the accelerator is *simultaneous with* the deflection of the needle.' Most ordinary sentences in English are A-series expressions, in that any verb will typically be *tensed*, i.e. its inflexion will indicate position in the A-series: 'Enid *is* dancing', 'Eric *wishes* to speak', 'Frank's performance *was* grotesque', 'Hermione *astounded* the spectators', 'Jeff *will be* here shortly.' When philosophers wish to use an expression that is truly *tenseless*, i.e. one that does not indicate position in the A-series, however vaguely, they often have to resort to rather artificial expressions, such as 'The Armistice OCCURS (tenselessly) on 11 November 1918', or 'The Sun BE 8 light-minutes from the earth.' Awkward though such expressions are, they can undoubtedly be useful in philosophical discussion, and in what follows capitals will be used to denote tenseless expressions.

Commenting on Broad's metaphor for the passage of time, we said that it represented time twice over: there was the order of the houses, *and* the movement of the beam. But now we have McTaggart's distinction between the A-series and the B-series at our disposal, we can see that perhaps this is not a defect in the model. For even if there is only one time, it does, it seems, have two aspects, captured by A-series and B-series expressions. Thus, the order of the houses, which does not change, can be said to represent the B-series. The movement of the light-beam, on the other hand, represents changing position in the A-series. Still, there is something odd about representing the A-series and B-series as two separate things. We do, admittedly, have two different kinds of expression here, but surely there is just *one* thing in the world to which they correspond. One way of putting this is that the A-series and B-series descriptions are not logically independent of one another. Expressions of one kind have implications for the truth or falsity of expressions of the other kind. Further, if there is a logical relationship between the A-series and the B-series, then we ought to be able to ask: which of them is the more fundamental? Which, of the two kinds of description, represents the more basic facts of reality? McTaggart's own answer is that, if one is more fundamental than the other, it must be the A-series that is the more fundamental. Part of his reason for thinking this is that change can only be understood in A-series terms—a suggestion we examine later in the chapter. Another consideration is this. Imagine a virtually omniscient being that has complete knowledge of the B-series position of every event in history. This being knows that a certain battle OCCURS before a certain coronation, which OCCURS before a certain revolution, and so on. Does this being, in virtue of possessing that knowledge, thereby know which of the various historical events is *present*? Does the being know, to put it in other words, what time it is now? The answer, surely, is no. Just from information about the B-series, however complete, it is not possible to deduce any A-series information of the kind 'The revolution is now over.' (It would, however, be possible to deduce a certain limited kind of A-series information. Thus from '*x* OCCURS before *y*' we can deduce 'If *y* is present, *x* is past.' But whether this is genuinely A-series information is a moot point.) And this is precisely what one would expect from the fact that whereas

A-series positions change, B-series positions do not. A changeless set of facts cannot imply a set of facts which holds for one time but not for other times. But now let us suppose instead that the being has instead complete knowledge of the *A-series* positions of events—knows, in other words, which events are present, which events past and future, and how distantly past and future they are. Could such a being, in virtue of possessing this complete A-series knowledge, deduce the B-series positions of events? Certainly. From the fact that the French Revolution is recently past, the Norman Conquest distantly past, the American Civil War present and the First World War future, the being can deduce that the Norman Conquest BE earlier than the French Revolution which BE earlier than the American Civil War which BE earlier than the First World War. B-series facts can be deduced from A-series facts, but not, it seems, vice versa. It would appear to follow that the A-series is the more fundamental.

However, there is a problem. For according to McTaggart, the A-series cannot exist, since it contains a contradiction. The contradiction arises through the mutual incompatibility of two undeniable propositions concerning the nature of the A-series. The first proposition is that different A-series positions are incompatible: if an event is present, it cannot be either past or future. The second proposition is that, given that time passes, every event must exhibit all A-series positions. If the A-series exists, both these propositions must be true. But, being inconsistent with each other, they cannot be true together. Thus, the A-series does not, indeed cannot, exist. The argument in full looks like this:

McTaggart's Proof of the Unreality of Time
1. If time is real, then there is an A-series (the A-series being the most fundamental kind of temporal series).
2. Different A-series positions are mutually incompatible, so no event can exhibit more than one of them.
3. If there is an A-series, then, since the A-series positions of events change, all events have all A-series positions.

Therefore:

4. If there is an A-series, then any event both has only one A-series position *and* has them all. But this is absurd.

Therefore:

5. There is no A-series.

Therefore:

6. Time is unreal.

At first sight, it seems that either premiss (2), or premiss (3), has been wilfully misinterpreted. The sense in which different A-series positions are incompatible (premiss (2)) is that no event can exhibit more than one of them *simultaneously*. And the sense in which all events have all A-series positions (premiss (3)) is that they exhibit those different positions *successively*: an event is *first* future, *then* present *then* past. So we only get a contradiction, it seems, if we either misread (2) as meaning that events cannot exhibit different A-series positions even successively, or misread (3) as meaning that all events have all A-series positions simultaneously.

Suppose, then, that we adopt the most natural reading of (3) and say that an event, such as Aunt Jane's visit, exhibits different A-series positions successively. What does this actually mean? 'Successively', surely, means 'at different times'. So an event is future at one time, present at another time and past at yet another time. But how do we express these times? We have two options: we can express them in A-series terms, or in B-series terms. (We assume here that there is no other kind of temporal expression, or at least not one that could help us here, since what we need is something that expresses *order*.) Suppose we express them in B-series terms. Then we get something like the following:

Aunt Jane's visit is: *in the distant future* on <u>Wednesday, 30 August 1837</u>
in the near future on <u>Sunday, 8 April 1962</u>
present on <u>Monday, 9 April 1962</u>
in the recent past on <u>Tuesday, 17 April 1962</u>
in the remote past on <u>Friday, 4 July 2099</u>

We have certainly avoided contradiction here, as being in the distant future on Wednesday, 30 August 1837 is perfectly compatible with being in the remote past on Friday, 4 July 2099. But the way in which we have avoided contradiction is by relativizing A-series positions (those in italics) to B-series positions (those underlined), and by relativizing those A-series positions we have, in effect, got rid of them. What we are left with is a series of pure B-series expressions, containing no A-series information at all. We can show this by the following tests: if the expressions above were genuine A-series expressions, then they should express changing states of affairs, and it should be possible to deduce which time was present. But they do not express changing states of affairs. If Aunt Jane's visit is present on 9 April 1962, then it remains present on that date for all time. The expression, in fact, conveys nothing more than that Aunt Jane's visit OCCURS on Monday, 9 April 1962. And we have no idea, from studying the list of propositions relating to her visit, whether Aunt Jane is still here, or due to arrive next week, or long since departed this world.

So, if we wish to preserve the reality of the A-series, it would be wise to interpret 'Events exhibit different A-series positions successively' as 'Events exhibit different A-series positions at different A-series times.' We now have something like the following:

Aunt Jane's visit is: *in the distant future* **in the remote past**
                        in the near future **in the recent past**
                        present **in the present**
                        in the recent past **in the near future**
                        in the remote past **in the distant future**

Again, contradiction is avoided, and this time not at the expense of losing the A-series. What we have done is to qualify A-series expressions by *second-order* A-series expressions (those in bold). These are genuinely A-series expressions, since they will cease to be true with the lapse of time, and also we can deduce from them that Aunt Jane is here now. Unfortunately, however, our description of Aunt Jane's visit is incomplete: it only captures the present state of affairs. But time consists of more than the present moment. There is also the past. Some would say, there is also the future, but perhaps this is more controversial. After all, when we were discussing Broad's metaphor, we suggested that perhaps there should be no houses that were about to be illuminated, as this would imply that the future existed in some sense. So let us say that the future is unreal, and we do not have to include, in our description of reality, any reference to future fact. Nevertheless, the reality of the past presses itself upon us. We do not have to exert ourselves much to recall what has just happened, and with a little more effort we can recall what happened some time ago. We are also surrounded by traces of what happened years, even centuries ago. Can I roam around a ruined castle, visit a museum, look at a photograph album or examine the rings on a tree stump, without being aware of the reality of the past? Well then, let us say that the past is real—as real as the present, only it just does not happen to be where (in time) we are. It follows then there are, in reality, past facts. What facts are these? Well, if it is a present fact that Aunt Jane's visit is *present in the present*, then it is also a present fact, something which obtains now, that the visit is future in the past. The *past* fact is that it is (*is*, note) future in the *present*. So our corresponding collection of past facts looks like this:

Aunt Jane's visit is: *in the distant future* **in the remote past**
in the near future **in the present**
present **in the near future**
in the recent past **in the not-quite-so-near future**
in the remote past **in the distant future**

This list is perfectly consistent. It is, however, entirely *inconsistent* with the first list, the list of present facts. So, qualifying the first-order A-series expressions by second-order A-series expressions does not, after all, avoid contradiction.

Again, it is useful to consider an analogy with space. Suppose we were to treat *hereness* as something that attached to places whether or not anyone was in them. That is, suppose we were to say that whether or not a place was 'here' was *not* just a matter of one's spatial perspective. Then there would be a problem. For *hereness* is not just confined to a single place: I correctly judge that the place where I am standing is 'here', you correctly judge that the place you are standing is likewise 'here'. But you and I are not standing in the same place. So you would (again, correctly) judge that the place *I* was standing in was 'there', and I would (again, correctly) judge that the place *you* were standing in was 'there'. So each of these two distinct places is both 'here' and 'there': is this not a contradiction? And if it is, one might wonder why McTaggart did not produce an argument for the unreality of space on this basis. The reason he did not, and the reason why there is no serious threat of contradiction, is that, of course, 'here' and 'there' are clearly perspectival. A place is not here independently of someone's location, or spatial perspective. We can say all we want to about where things are in space in terms of the distance relations between those things and other things—in terms, that is, of what we might call a 'spatial B-series', because facts about the relations between places do not vary depending on one's own position in space. The temporal equivalent of this move was to relativize A-series positions to B-series positions. But we did not want to take that route (without further thought, at least), precisely because it required us to give up what we took to be distinctive about time, what distinguished time from space, namely the fact that presentness, unlike hereness, is *not* perspectival. An event is present *full stop,* we wanted to say, not merely present with respect to a time, or someone's temporal perspective. But this is precisely what led to contradiction, so what are we to do?

First Response: Presentism

Since contradiction loomed precisely at the point at which we introduced past facts, the first, obvious but radical, response is *not* to introduce past facts. In other words, we simply deny the reality of the past, and refuse to countenance past facts. The only facts that there are, are present facts, and no list of purely present facts will contain a contradiction. This does not require us to abstain from uttering any past-tense statement, for when we say that Aunt Jane's visit *was* future, we are not, appearances aside, alluding to a past fact, but expressing a rather complex present fact. It is because Aunt Jane's visit is *now* present that it *was* future. This move, of restricting reality to the present moment, is known as *presentism*. Presentism is, perhaps, our intuitive view of time. For, just as we suppose that the future is not yet real, a natural way of thinking of the past is that it has *ceased* to be real. It is not that past events are *still there,* in some sense, like Broad's houses that have already been illuminated; rather, they have suffered a drastic loss of status. But, even if presentism is our intuitive view, or a slightly more precise version of it, two questions remain: first, what is it exactly to say that the past (or future) is unreal? Secondly, if the past is unreal, what makes true our statements, beliefs, or memories concerning the past?

What is it to say that anything is unreal? There are two quite different things we might want to indicate when we talk of the unreality of something. Consider first the case of a purely fictional character, such as Bradley Headstone (in Dickens's *Our Mutual Friend*). Although there are things we can truly say of him, such as that he is a schoolmaster, or that he is consumed with hatred and envy whenever he thinks of Eugene Wrayburn's liaison with Lizzie Hexham, Headstone does not exist in reality. No actual being corresponds (let us hope) to Dickens's description of Bradley Headstone. But we do not contradict ourselves when, at one and the same time, we concede his unreality, *and* continue to utter what we take to be truths about him, for such utterances are, so it may plausibly be argued, just shorthand for 'In the fiction *Our Mutual Friend,* it is true that . . .' or 'It is fictionally true that . . .'. But now consider a rather different case, a saying attributed to Democritus: 'By convention sweet, by convention bitter, by convention hot, by convention cold, by convention colour: but in reality atoms and void' (Kirk, Raven, and Schofield 1983, 410), Here, properties such as colour, warmth, and sweetness are not being dismissed *as fictions,* exactly. But what accounts for the fact that we perceive the world as having these properties is the behaviour of atoms in the void: this is the most fundamental level of reality. Exactly this distinction between perceived properties and the fundamental properties of things turns up again in later philosophers, such as Descartes and John Locke (1632–1704). One way we can express the distinction is this: there can be true statements containing colour terms (or terms for warmth, sweetness, etc.) but what makes these statements true are not facts about colour, but rather facts about the atomic structure of the objects to which we attribute colour. It is in this sense that colours are 'not real', and it is a quite different sense from the one in which Bradley Headstone is not real. 'My motor car is bright yellow' is, if true, true without qualification. It is not shorthand (at least, in normal circumstances) for '*It is fictionally true* that my motor car is bright yellow.'

So, let us turn back to the question of the unreality of the past. I doubt if many people would be attracted to the view that the past is unreal in the sense of being a fiction. Not all my memories are the product of my imagination. That leaves us with the second sense: statements about the past can be true, and indeed true without qualification, but what makes them true is not *past* fact. If the past is unreal, there are no past facts—i.e. past bits of reality—to make true anything at all. So what does make true statements about the past— for example, the statement that Colonel Rathbone was shot in the conservatory? Presumably, something that belongs to reality, and since the presentist has restricted reality to what is present, the facts that make statements about the past true must be *present* facts. What present facts, exactly? It would be evading the issue simply to say, for example, 'The fact that Colonel Rathbone was shot in the conservatory', because this is either an illegitimate reference to past fact, or simply a way of restating precisely what needs explaining, namely the truth about Colonel Rathbone. What else can we appeal to? Well, the past leaves its traces upon the present: creatures long extinct leave their fossils, remote civilizations leave memorials in the form of castles, pyramids, tombs, and temples, what happened three summers ago is still to be found in our memory. In the case of the late Colonel, we can point to the bloodstains on the tiles, the tell-tale bullet holes through the watering-can, the enigmatic marks carved on the cactus, by which (we conjecture) the Colonel hero-ically attempted to identify his assassin. Ideally, the sum total of truths about the present would entail a complete set of truths about the past: given what is the case *now*, such-and-such must have been the case *then*. But does present truth entail anything about the past? Let us say that there are any number of ways in which history could have led up to precisely the point we have reached now. This, on a large scale, mirrors what we are inclined to think of individual events: this explosion, or at least this kind of explosion, could have been caused by a lightning strike, or by a chemical leakage, or an electrical short-circuit, or . . . . On this picture, many different pasts are compatible with the present state of the world (Figure 21). If this is the correct picture, then presentism, at least as we have portrayed it, is in trouble. For if any number of pasts are compatible with the present state of affairs, and it is only the present state of affairs that can make true or false statements about the past, then no statement about the past is either true or false. What is now the case, as we might put it, *underdetermines* what was the case. So, to guarantee a definite truth-value to every statement we might make about the past (i.e. to guarantee that every such statement will be determin-

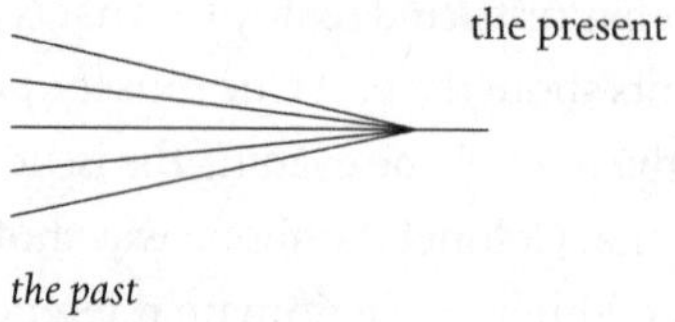

Fig. 21. **The past as a set of possibilities**

ately true or determinately false), the presentist has to assume that only one past is compatible with the present state of the world: only one course of history could possibly have led up to this point.

We have given an account of the mechanism on which the presentist must rely in order to account for the truth or falsehood of statements about the past. It is a causal mechanism linking events or states at different times. But is the presentist capable of articulating that mechanism? The question is a pertinent one, for, in order to be able to exploit the mechanism, it must be possible for the presentist to make true statements *about* that mechanism, statements that do not violate presentist assumptions. But this is where the presentist comes unstuck. Consider the statement 'The past leaves causal traces on the present.' What, according to the presentist, makes this statement true? Well, present fact, presumably, since that is the only kind of fact available. But what purely present fact could make true a statement about the causal relations between different times? We can make sense of a past event leaving its causal traces on the present (last night's wild party has left a number of traces around my sitting-room, for instance: the smashed wineglass, the shoe-marks on the piano, the underwear draped over the sofa), but can we make sense of the *causal relation* between that event and the present traces *itself* leaving its traces on the present? The idea is a distinctly odd one. Any statement about the relation between different times (or between the events that occurred at those times) requires us to stand, in thought at least, outside those times and view them as of equal status. There cannot be a relation if one of the things the relation is supposed to relate is just not part of reality. It looks, then, as if the presentist is not entitled to assume the only mechanism that can explain, in terms of present fact, how statements about the past can be true.

Someone will object that we have foisted on the presentists an account that they are not obliged to buy. Why should they have to explain the truth of statements about the past in terms of present fact? Why can they not say, simply, that such-and-such was the case?

Well, they could, but it then becomes unclear what they mean when they deny the reality of the past, unless this amounts to no more than the trivial truth that the past is not part of *present* reality. In any case, even if the mechanism that allows statements about the past to be true does not involve causal relations between times, it would still be an unfortunate result if presentism turned out to be incompatible with the intuitive view that causal relations do link events occurring at different times. But as was pointed out above, the causal connection between the past and the present seems to require that the past is as real as the present.

These worries are, at least, enough for it to be wise to consider an alternative response to McTaggart's argument.

Second Response: The B-Theory

One of the premisses of McTaggart's argument is that the A-series is more fundamental than the B-series. To put it another way, B-series facts obtain because A-series facts do. To put it yet another way, the B-series does not, in one sense, exist, at least as an independent series, but B-series statements are true in virtue of there being A-series facts. Thus, the fact that *a* is present and *b* is present makes true, on a particular occasion, the statement that *a* is simultaneous with *b*. But let us see what happens if we reject that premiss, or rather if we do more than just reject it, but turn it on its head. What if, in other words, we say that there is only a *B-series*, events have only B-series locations, and there are only B-series facts? Let us call this idea the supposition that the universe is a *B-universe*, and proponents of such a supposition 'B-theorists'. (Similarly, an *A-universe* is one that contains an A-series, and champions of the A-universe *A-theorists*.)

An advantage of this move of denying the reality of the A-series but asserting the reality of the B-series is that it enables us to dispose of McTaggart's contradiction very simply. Events are not, without qualification, past, present, or future, but rather past, etc. in relation to a time, specifiable in B-series terms. Thus Aunt Jane's visit is not simply present, but rather present on 9 April 1962, which is to say nothing more than that it OCCURS on 9 April 1962. And it is not simply past, but *past on* (e.g.) 17 April 1962 (and thereafter), which is to say no more than that it OCCURS earlier than 17 April 1962. So there is a perfectly straightforward sense in which all events exhibit all A-series positions, and that is that they exhibit what would otherwise be incompatible positions only in relation to different B-series locations. But there are some formidable obstacles to overcome before we adopt this solution. First, how do we account for the truth of A-series statements in a B-universe? For it would surely be unacceptable to say that harmless statements such as 'The post has just arrived' could not be true, in virtue of the fact that there is no such thing as the present moment. But it is not immediately obvious how A-series truths can be derived from B-series facts, given what was observed above, namely that even an omniscient (or nearly omniscient) being who had knowledge of all B-series facts would still be unable to deduce what time it was *now*. Secondly, since the passage of time requires an A-series (for the passage of time just is change in A-series facts), it follows that time does not pass in a B-universe. So surely there can be no question of our actually living in a B-universe, for nothing is more obviously a feature of ordinary experience than time's passage. Thirdly, and relatedly, how can anything change in a B-universe? For surely nothing can change, by for example becoming colder, or larger, or less dense, or moving about, unless time itself moves on so that earlier states are replaced by later ones?

In answer to the first concern, we could point out that not knowing what time it is now does not represent a gap in our omniscient being's knowledge, for, this being a B-universe, no time is present: this would be an A-series fact, and there are no such facts. But we have to say more than that, for there are countless statements made every day, not to mention unspoken beliefs, concerning what is going on now, and it would be absurd to say that none of them was capable of being true. The B-theorist needs to say something like this. What determines whether a statement such as 'The meteor shower is now visible' is true are two things: first, when in the B-series the event in question takes place; secondly, when in the B-series the statement is made. So, let us suppose that the meteor shower BE visible (from a certain position on the surface of the Earth) on 15 November 2001; and let us further suppose that the statement that it is visible also BE made on 15 November 2001. Then the statement is true. To take another example, suppose we say, sometime after 1918, 'The Great War is over.' What we say is true, since the Great War ended in 1918. This account of the truth of A-series statements precisely mirrors the most natural account of the truth of statements like 'The treasure is buried here.' If the treasure is buried at place X (as it is customarily described on a map), and we make our statement at X, our statement is true. So no esoteric property of hereness is required for the truth of such statements. As we noted earlier, what place is correctly described as 'here' depends where we are in space: it simply reflects our spatial perspective on the world. Similarly, for the B-theorist, what time or event is correctly described as 'present' depends where we are in time: it simply reflects our temporal perspective on the world.

But that now raises a tricky question: what, in the B-universe, determines my temporal perspective? For although I can move about in space almost at will, thus determining my own spatial perspective, I have no choice over where I am in time. Something other than me places me at a time. Moreover, what time I am is something that *changes,* and again I have no control over that. So this is one illustration of the second and third concerns, for having a temporal perspective on time is an essential aspect of my experience of time, and the fact that I have to have the perspective I do, and that it changes, seems only explicable if we live in an *A-universe* (i.e. one that has a real A-series). For it is hard to resist the thought that we occupy the perspective we do because we are necessarily located at the *present* time, and our perspective shifts with the shifting present. This is, perhaps, the hardest problem for the B-theorist to deal with, raising the tricky issue of our identity as persons through time. (This is a problem that we will address again in 'Concluding Thoughts'.)

We can, however, say something about the third issue here. Change in the B-universe just *is* the fact that objects have different properties at different times. Thus, I am 6 foot tall in 2001, but only 3 foot tall in 1966. In an A-universe, in contrast, the change in my height is itself something that undergoes change: it is first present and then increasingly past. But why, the B-universe may insist, should changes themselves have to change? There is just one kind of change, namely variation in properties over time. But for this to work, it must be the same object that first has one property and then, at a later time, has another property. And we have just remarked that it is by no means obvious that things can change their locations in a B-universe. So the problem of change and the problem of changing perspective need to be solved together.

Why Is There Only One Present?

We seem to be in a rather unhappy situation. For both of the two possible means of avoiding McTaggart's contradiction are beset by problems. The presentist preserves the A-series (or part of it), but seems unable to account for the truth of statements about the past, whilst the proponent of the B-universe seems unable to give a satisfactory account of change. While we are contemplating this fraught matter, let us end the chapter with another conundrum.

'There's only one now', yells the advertisement, exhorting one to take action *without delay* and buy a new motor car, dishwasher, fitted kitchen, life-insurance policy. Undoubtedly, there is only one now, but is this just a truism, or a substantial and important fact about the world? And if it is a substantial fact, what explains it? Could things have been otherwise? Could there have been *more* than one now? For the B-theorist, at least, there is no problem here. Every time I utter the word 'now', I pick out the time at which I am speaking, just as every time I utter the word 'here' I pick out the place from which I am speaking. 'Now' is simply an *indexical* term, that is, its reference varies in systematic ways with the context in which it is used. So it is really just a truism that there is only one now, for any particular use of the term 'now'. But in another sense, of course, there are many 'nows', for people situated differently in time will refer to different times by the word 'now'. In just this sense, there is more than one 'me', as 'me' just refers to the person using the term, and it can be used by different people. But when I say, 'There is only one me', what I say is true, by virtue of the fact that this particular use of the term 'me' picks out exactly one person. What I express could equally well have been expressed by 'There is only one Robin Le Poidevin' (where the statement is intended to refer to a particular person, and not to make the assertion that only one person is *called* 'Robin Le Poidevin').

But for the A-universe proponent, 'There is only one now' is not just a truism, and the meaning of the word 'now' is not captured, or at least not exhaustively captured, by treating it as an indexical term on a par with 'here' and 'me'. For the presentist, of course, the word 'now' has a special meaning: it captures the limits of reality. Everything that is real exists now. Similarly, for someone who thought that the past is real, but the future not, 'now' conveys something special: it is the leading edge of reality, the latest point in history to come into being. And even for the A-universe proponent who thinks of past, present, and future as being equally real, the now still has a unique status: it is where things happen, the shifting boundary between past and future. 'There is only one now' conveys, for the A-theorist, this uniqueness of the present moment, and could indeed be adopted as a defining slogan by champions of the A-universe, as well as by advertisers of domestic luxuries. Are we then not entitled to ask *why* the now, or present moment, is unique? That is, not what it means to say that it is unique—we have already attempted to answer that question—but what other facts we can appeal to in order to understand why it has turned out that there is just the one now, and not several. We, or at least the A-theorists, have to tread carefully here. If the explanation is a purely logical one, so that it emerges that it would actually be self-contradictory to say 'There is no now' or 'There is more than one now', then we run the risk of reducing our slogan to the status of a mere truism. For example, suppose we point out that the 'is' in 'There is only one now' is present tensed, referring to what exists now, and that, by definition, only one now exists *now*. That would certainly reduce the slogan to triviality, and so should be avoided by the A-theorist.

At first sight, it might seem that the presentist is in the best position to explain the uniqueness of the now. For only the present— what exists now—is real, and there is only one reality. This may explain why there is *at least* one now, but it does not actually follow from this that there is *only* one now. It is consistent with the propositions that only the now is real and that as there is only one reality that reality should contain more than one now. What is ruled out is that it should contain anything that is *not* now. So presentism is not enough to guarantee the uniqueness of the now.

Here is another, more promising, explanation. Suppose that there were two nows. This must surely imply that two *different* times were now, for only then would the nows be distinct. But then, if two different times were both now, then they would be simultaneous. This depends on a feature of the A-universe pointed out above, that B-series facts derive from A-series facts. So, for example, if *a* and *b* are both present, then *a* is simultaneous with *b*. But two distinct times are, by definition, *not* simultaneous with each other: one must be earlier or later than the other. So there cannot be two or more nows. (We would still need, however, to find another explanation for why there is *at least* one now.) What is suspect about this explanation is that it appeals to a B-series relation, namely simultaneity. For according to the A-theorist, the A-series is *more fundamental* than the B-series, so it should not be necessary to appeal to the B-series at all in explaining some fundamental feature of the A-series.

Finally, we could avoid all talk of the B-series by pointing out that what makes individual times the times they are and distinct from other times is position in the A-series. So it is just part of the definition of the A-series that there is only one present: that is, there is only one such position in the A-series. But still this does not give us a complete explanation, for we can go on to ask 'And why is there only one A-series?' It seems, then, that it is no easy feat to explain, without assuming that we in fact live in a B-universe, why there is only one now.

McTaggart's distinction between the A-series and the B-series, and his famous proof of the unreality of time, provides an extraordinarily rich field for thought, and much has been written on the subject. So it is sad to report that the rest of his writings have been relatively neglected, and that his own intellectual influence, once so powerful over such significant thinkers as Russell and Moore, dwindled in the last two decades of his life. Ernest Rutherford, who was later to occupy the Cavendish Chair of Physics at Cambridge, records briefly in his diary his first meeting, as a young man, with McTaggart over one of the latter's notoriously frugal breakfasts. The occasion does not appear to impressed him greatly: '*May 1896.* Breakfast with McTaggart, Hegelian philosopher and Fellow of Trinity, but he gave me a very poor breakfast worse luck,' he writes. 'His philosophy doesn't count for much when brought face to face with two kidneys, a thing I abhor.'

Questions

Is 'One second per second' a satisfactory answer to the question 'How fast does time pass?' Is it a legitimate question to ask in the first place?

I can know what time it is now only by being located at a particular position in time. Does that mean that a timeless being could not know what the time was? If so, does this show that there must be something wrong with the idea of God as *both* timeless *and* omniscient?

Is there only one now? Why?

could not know what the time was? If so, does this show that there must be something wrong with the idea of God as *both* timeless *and* omniscient?

Is there only one now? Why?

CHAPTER 9

The Cinematic Universe

... history is a pattern
Of timeless moments.

T. S. Eliot, 'Little Gidding'

Muybridge's Horse and Zeno's Arrow

Has photography changed the way we think about time? More particularly, has it affected our view of change and motion? Early photographic processes involved long exposure times, and so the first photographs tended to be still lifes, such as Fox Talbot's 'The Open Door', which depicted a broom leaning against a doorway at his home, Lacock Abbey in Wiltshire. But, as the technique developed, photographs gradually became 'snapshots'. By the 1870s, cameras were capable of taking photographs with exposure times of a thousandth of a second or less. This allowed the American photographer Eadweard Muybridge to undertake a series of studies of animal movement. (His original name was Edward Muggeridge, but he later adopted an 'Anglo-Saxon' spelling.) The most famous of these, published in his *Human and Animal Locomotion* in 1878, was a series of stills of a galloping horse, Occident, who belonged to the one-time Governor of California, Leland Stanford. Muybridge succeeded in capturing on film the different positions of Occident's legs during the gallop, thus showing for the first time that the conventional representation in paintings of galloping horses, with all four legs off the ground, was mistaken. Similar studies of locomotion were undertaken around the same time by the French physiologist Etienne Marey, who described his technique as '*chronophotographie*'. Photography had reduced motion to a series of stills—images in which the object is frozen in the posture it adopts at an instant of time. Continuous motion is lost, although, as the inventors of cinematography discovered, the experience of such motion may be induced by projecting those stills in rapid succession onto a screen.

That 'real' motion is similarly an illusion is the conclusion of the third of Zeno's paradoxes, known as the *Arrow*. The Arrow is quite different from the other paradoxes of Zeno we have discussed. For one thing, it does not, at least very evidently, invoke the notion of infinite divisibility. In fact, there is, as we shall see, a closer connection with the topic of the previous chapter, on the notion of the present, than with infinity. The gist of the paradox is this: when we look carefully at what motion is, we find just a series of states in which the moving object simply occupies a position in space. None of these states individually counts as motion, and yet, when we have described each state, there is nothing left over to describe about the motion. Motion itself thus seems to have disappeared. Zeno's descriptive deconstruction of motion is like Muybridge's photographic deconstruction: both present us with an idea of motion that is quite unlike our ordinary experience of it, and we realize that the way the world appears to us may be no less a product of our minds than the motion of figures on the cinema screen. But before giving in to such an assault on our ordinary beliefs, however, we should look in detail at the structure of Zeno's paradox.

No Motion at an Instant?

The Arrow can be presented in different ways, and we shall look at no less than three versions of it in this chapter. What was Zeno's original conception? Here, again, we have to rely on later commentators, perhaps the most helpful being Simplicius. This is his account of the paradox:

> The flying missile occupies a space equal to itself at each instant, and so during the whole time of its flight; what occupies a space equal to itself at an instant is not in motion, since nothing is in motion at an instant; but what is not in motion is at rest, since everything is either in motion or at rest: therefore the flying missile, while it is in flight, is at rest during the whole time of its flight. (Lee 1936, 53)

For 'missile' we will substitute, as everyone does, an arrow. What is meant by 'occupies a space equal to itself at each instant'? Presumably, that the arrow fills a volume of space equal to its own volume. How, we wonder, could any object do otherwise, whether moving or not? Perhaps the idea is this. We could think of the arrow marking out a region of space as it moves, rather like the vapour trail behind an aeroplane. The region of space thus marked out will be greater than the volume of the arrow. But, in an instant, an object can only mark out a region of space equal to its own volume, and so cannot be moving in that instant. (But is it therefore not moving *at* an instant? We take a look at the distinction between 'in' and 'at' an instant in the next section.)

Let us set the argument out more explicitly. It looks somewhat like this:

The Arrow: first version
1. If the arrow moves throughout the period of its flight, then it moves at each instant of that period.
2. The arrow occupies a space equal to its own volume at each instant.

3. If the arrow occupies a space equal to its own volume at an instant, then it is not in motion at that instant.

Therefore (from 2 and 3):

4. The arrow is not in motion at any instant of the period.

Therefore (from 1 and 4):

5. The arrow does not move throughout the period of its flight.

Everything depends here on what we mean by 'instant'. For premiss (2) to be at all plausible, we need to interpret this as an *indivisible* point of time, not further resolvable into smaller items. If there can be nothing smaller than an instant, then whatever takes place at an instant cannot be differentiated into distinct states obtaining in smaller moments. It is because the instant is indivisible that the arrow cannot move at an instant, for motion involves occupying different positions. So any time at which the arrow is moving is divisible into distinct parts: the part where the arrow occupies just *this* position, the part where it occupies just *that* position, and so on. But an instant is not so divisible.

As a way of making (4) seem more plausible, imagine some omnipotent deity who decides to bring a world into being—but only for a single instant. Can there be change in such a world? Surely not, for there would be no time in which change could occur.

Aristotle, who discusses the Arrow along with Zeno's other paradoxes of motion in the *Physics,* has this, rather terse, objection: '[the conclusion] follows from the assumption that time is composed of moments: if this assumption is not granted, the conclusion will not follow' (*Physics* 239ᵇ30–3). The premiss this observation is most relevant to is (1), since one could defend (1) on the basis that there is nothing to a period other than the instants within it, and what is true of the period must therefore be true of every instant within it. But this defence is suspect whether or not we suppose time to be composed of instants. For what is true of the whole may not be true of its parts. For example, this wardrobe is 7 foot tall, but none of the individual pieces of wood from which it is made is 7 foot tall. Or, to take a temporal case, the concerto lasted 40 minutes, but none of the movements which comprise the concerto lasted that long. Perhaps, then, we should present (1) as an application of the following principle:

If an object has a certain property throughout a period of time then that object has that property at every instant of the period.

Now, clearly, for most properties, this principle is perfectly acceptable: green, cubic, composed of copper, at 10 °C, being someone's uncle, being 2 miles from Puddingdale, reflecting light, being thought about, etc. (1), then, will only be vulnerable if there is any good reason to doubt that the general principle above applies to the property of *being in motion*.

Why does Aristotle suppose the Arrow to depend on the premiss that time is composed of instants? As we have just seen, it is possible to defend (1) without recourse to such an assumption, and indeed explicit appeal to it would invite the charge of fallacy. Now it may be that Aristotle takes Zeno to be making a fairly obvious blunder, but there is a more interesting possibility.

One rather obvious assumption of the Arrow is that there are such things as instants, not simply as a useful idea, but as independently existing entities, for there to be truths about what is the case at those instants. But this assumption may have some rather surprising consequences. Let us imagine that the indivisible instants of time at which the Arrow occupies a space just its own size have a small, but non-zero duration. Then periods of time would be composed of such 'time atoms', an idea we first encountered in Chapter 7, and the number of such atoms in a period would determine its length (assuming that each atom has the same duration). Time atoms would be the fundamental unit from which periods were built up. Time atoms just are instants, so this picture of time raises no difficulties for the Arrow's assumption that there are instants. But now suppose instead that time is not discrete (composed of time atoms) but continuous, so that each period of time were indefinitely divisible. What implications would this have for instants? Although time is continuous, we can attempt to avoid the consequence that would appear to flow from this, namely that each period contains an infinite number of instants, by the assumption of *finitism*. Nothing is *actually* infinite (there is no infinitely large object, for example), but only *potentially* infinite (see Chapters 6 and 7 for a discussion of this distinction). So a length is infinitely divisible only in the sense that, however many divisions you have made, you can always make more. The process of dividing has no limit. But the divisions do not exist independently of our making them, so the potentially infinite divisibility of a length does not imply the actual existence of an infinite number of divisions within that length. And as with length, so with time: a period of time is infinitely divisible, but only potentially so. Durationless instants do not exist independently of their being marked in some way. This view of the divisibility of space and time, as we saw in Chapter 7, provides Aristotle with an answer to two other Zenonian paradoxes of motion, the Dichotomy and the Achilles. Both paradoxes start from the assumption of the continuity of time and space, and derive the unpalatable conclusion that motion involves achieving the impossible: the traversal of an infinite number of subdistances in a finite length of time. Aristotle dissolves the paradoxes by denying that divisions have any independent existence. Admittedly, Aristotle only explicitly applies this to spatial divisions, but the same reasoning leads to a similar conclusion concerning temporal divisions.

This leads us to the following conjecture: Aristotle, quite correctly, takes the Arrow paradox to depend on an assumption about instants: that they really exist. But he takes this assumption to imply the existence of time atoms, i.e. tiny intervals of time that have non-zero duration but are nevertheless indivisible, which he expresses as the view that time is composed of instants. So attributing to Zeno the premiss that time is composed of instants is simply an expression of the fact that the Arrow depends on the real existence of instants. Given that time is continuous, however, instants do not have the independent existence required for the Arrow to work.

Whether or not this reconstruction of Aristotle's thinking holds water, it is certainly true that a number of commentators have attributed an atomistic premiss to Zeno. There is something attractive about the suggestion. For one thing, it presents a pleasingly symmetrical picture of the overall dialectic of Zeno's four famous paradoxes of motion. The Dichotomy and Achilles show that motion is impossible if space and time are continuous; the Arrow and Stadium show that motion is impossible if space and time are discrete. Since motion is impossible without time, and time must either have a discrete or continuous structure, motion is impossible.

Well that is just a suggestion. I do not press it too strongly, as it could be argued that Aristotle's conception of the potential–actual infinite distinction only conflicts with infinite numbers of spatial points, not infinite numbers of instants. Given Aristotle's conception of the actual infinite as that which exists *all at once,* a continuum of real spatial points would count as an actual infinite by this definition, whereas a continuum of temporal points would not, as the points are successive and not simultaneous. So perhaps Aristotle's finitism would allow the existence of instants even if time were infinitely divisible. What really matters is that an instant is indivisible, whether or not it is conceived as having duration (as in the picture of time as composed of time atoms) or durationless. It is because an instant is indivisible that there can be no motion within it.

There is another problem, however: why should we take the existence of indivisible instants seriously, when we can say everything we want to say about time by talking only of intervals (including, where necessary, infinitely small intervals)? Well, there is one instant at least whose existence is hard to impugn: the present moment. In the next section, we will see what happens to Zeno's Arrow when we build reference to the present into it.

Perhaps the most effective answer to the first version of the Arrow is to produce an account of motion that is both plausible and which undermines at least one of the premisses. So what is motion? The obvious answer defines motion as follows: motion consists just in an object's occupying different places at different times. After all, is this not exactly what Muybridge's photographic study of animal locomotion revealed? This approach to motion we will call, following Bertrand Russell, the *static account of motion*. Now the Arrow seems to depend on this view of motion, implying as it does that an object can only occupy a position in space at any one instant. But, ironically, the static account of motion also appears to undermine the Arrow. There are two ways in which one might expand on this point. One is to say that the static account falsifies premiss (1). We could concede that there is no such thing as motion (or indeed change in general) at an instant, but merely the occupancy of a particular position (or state), and insist that motion is attributable to an object only over a period of time. By analogy, an extended object may be 10 foot wide without it being true that it is 10 foot wide at every indivisible spatial point (what would it mean to say that it is 10 foot wide *at a point*?). This is Russell's response to the Arrow (Russell 1903, 467–73). It is a rather surprising response, however, since there is a kind of motion that seemingly obliges us to talk of motion at an instant, namely *acceleration*. If an object accelerates continuously through a period, it surely has a different velocity at each successive instant. One could, perhaps, insist that this is merely a theoretical abstraction, but it would have to go hand in hand with the view that instants themselves are theoretical abstractions. A much more plausible approach takes our analysis of motion to undermine premiss (3). There is no need to deny that objects can move at an instant, but they do so only in a derivative sense. An object is in motion at an instant if (and only if) it occupies different positions at times immediately preceding and/or immediately succeeding that instant. What is true at an instant thus depends in part on what is true at *other* times. So from the indisputable fact that an object necessarily occupies a space just its own size at an instant it does not follow that an object cannot be in motion at an instant. By analogy, the shape of an object depends on its individual constituents, but they are not required to have the same shape as the whole object. The full account of motion, then, looks like this:

The static account of motion. An object is in motion throughout a period if and only if the object occupies different positions at every instant of that period; it is in motion at an instant if and only if it occupies a different position at instants immediately before and after that instant.

The game is not yet up for Zeno, however, for we can provide another reconstruction of the Arrow, one that brings out the limitations of the static account of motion.

No Motion in the Present?

When Aristotle discusses the Arrow, he uses the phrase εν το νυν. This can be translated either as 'in an instant' or 'in the *now*'. The second translation opens up the interesting possibility that it is the *present* moment and not merely some arbitrary instant, that is the key idea in the Arrow. Now, assuming time to be infinitely divisible, the present can have no duration at all, for if it did, we could divide it into parts, and some parts would be earlier than others. But something that is present cannot be earlier than anything else that is also present! So the present cannot have earlier and later parts, which is to say that it can have no duration. So, using the slightly more idiomatic phrase 'in the present' for 'in the now', substituting this phrase in place of 'at an instant' in the first version of the Arrow, and simplifying the argument somewhat, we obtain the following:

The Arrow: second version
1. If the arrow moves throughout the period of its flight, then, when it moves, it moves in the present.
2. The arrow is not in motion in the present.
   Therefore:
3. The arrow does not move throughout the period of its flight.

Recall one objection to the first reconstruction of the Arrow: that we could avoid talk of indivisible instants altogether in favour of intervals, and in so doing make irrelevant the alleged fact that nothing moves at an instant. But turning the focus of the argument onto the present moment explains why the moment in question must be an indivisible instant and not a period. For if the present were divisible into different parts, some would be earlier than others, and so not present. But every part of the present must itself be present, which is to say, of course, that the present has no earlier and later parts.

Another feature of this version is that it talks, not of what is happening *at* the present, but *in* the present (this being the literal translation of εν το νυν). This certainly makes the premises somewhat more plausible, for we could happily concede that nothing moves in (i.e. within the space of) the present. It is not immediately obvious how this helps, however, because we could still insist that the crucial question is what is true of the arrow *at* an instant, and the static account of motion allows us to say of the arrow that it is moving at an instant of time. It moves *at* an instant by virtue of that instant being part of a period *in* which the arrow is continuously in motion.

Nevertheless, the second version does represent a significant advance on the first. The most powerful objection to the first reconstruction was provided by the static analysis of motion: something moves in an instant by virtue of its position both at that instant and at other times. So talk of motion at an instant is derivative: its truth depends on what is happening over a period of time. It is this move that is challenged by the second version of Zeno's paradox. For what is true in the present should not be *derivative*, but *fundamental*. It is the privileged status of the present that insulates present fact from past and future fact. The static analysis of motion makes expressions like 'moves in the present' temporally hybrid, turning what purports to be a simple statement about the present into a complex statement about past, present, and future. But, we may imagine the champions of the present arguing, 'the arrow moves in the present' is a simple statement about the present, and should not be taken as elliptical for something else.

But in what sense is the present special? And should its special status imply anything about the nature of change and motion? Before answering these questions, however, we should consider whether there is a plausible alternative to the static account of motion, one that does justice to the intuition that 'the arrow moves in the present' is a simple assertion about what is presently the case. The static account of motion, in a nutshell, goes as follows: An object moves at a time by virtue of its position at that time *and* its position(s) at other times. The contradictory of this should be something like the following:

> *The dynamic account of motion*: an object's motion at a time is independent of the object's position at other times.

Perhaps so bald a statement does not merit the title of 'account'. Something, certainly, needs to be said to make it intelligible how the motion of an object at a time can be independent of what is happening at other times. Here are two ways we could spell out what is going on (are there any others?):

(i) It is an intrinsic property of an object that it is in motion at a particular time, the property in question being a disposition of the object to be elsewhere than the place it is. (An *intrinsic property* of an object being one that the object has independently of the existence or properties of other objects.)

(ii) Events, including those involving motion, are primitive, not decomposable into series of states. Similarly, intervals are primitive, not decomposable into series of instants. So talk of 'motion at a time' must always be interpreted as motion in an interval, however small.

Account (i) falls victim to the following counterexample: a stationary object is struck by a rapidly moving one, after which it begins to move. Though the object is *disposed* to be elsewhere at the moment of impact, as a result of its being subjected to a force at that moment, we would count that moment as the last moment of rest, rather than the first moment of motion. (i) would also conflict with the (more contentious) relationist assertion that constant motion is always relative to some other object or objects.

We are left, then, with (ii). It certainly undermines the first version of the Arrow, because it gives us a reason to reject the first premiss. Does it undermine the second? Let us reserve judgement for the time being, and return to the questions we raised earlier. What is it about the present that is special? And what effect does it have on our understanding of motion?

Zeno and the Presentist

In the previous chapter, we introduced two different conceptions of time and the universe. One conceives of the universe as an *A-universe*, namely one in which time passes, and in which A-series facts (that, e.g. the party is going on now) are more fundamental than B-series facts (that the party OCCURS later than the Ascot races). The other conceives of the universe as a *B-universe,* namely one in which time does not pass, and in which there exist only B-series facts, these being the facts that make A-series statements ('Claudius ruled the Roman Empire two millennia ago') true. Now, clearly it is on the A-universe view that the present is special, indeed unique. The B-universe view simply relegates terms like 'now' and 'present' to the status of indexical terms like 'here' and 'I'. Unfortunately, the A-universe view, as we saw, is threatened by McTaggart's paradox, and the obvious (perhaps the only) way out for the A-universe proponent is to adopt presentism, the view that only what is present is real. One way of expressing presentism is to say that all facts are present facts. This strongly suggests that what *was* and *will be* the case obtains only by virtue of what is now the case. (Suggests, but perhaps does not entail. On the other hand, what other account could the presentist give of what makes true statements about past and future? At any rate, in what follows, I shall take 'presentism' to refer to the theory that holds present fact to be the truth-makers of statements about past and future.)

Presentism, then, provides an account of what makes the present special, and indeed explains why motion, when it occurs, should occur in the present. For every aspect of reality is necessarily an aspect of *present* reality, so if motion is real, it must take place in the present. This now may help to explain what is suspect about the static account of motion, for if motion is pure and simply a present fact, it cannot depend on what is happening at *other* times. We have to be careful here, however. For presentism, although it rejects past and future fact, conceived of as parts of reality, allows for the truth of statements about past and future. (Assuming, of course, that the worries raised in the last chapter can be met, and the presentist can give a consistent account of the mechanism whereby statements about past and future can be made true by present fact.) The presentist, then, can allow that the arrow was in a different position from the one it now occupies, but has to insist that this is made true by present fact. So one set of present facts makes true 'the arrow was at s_1', another set makes true 'the arrow is at s_2', and yet another set 'the arrow will be at s_3', where s_1, s_2, and s_3 name different locations in space. So it seems, after all, as if present fact can, in principle, make it true that the arrow is moving, even when we understand motion in terms of the static analysis.

But there is a remaining problem for any attempt to reconcile presentism with the static analysis of motion, and it has to do with the extent to which the present is capable of making determinate past states of affairs. As we saw in the last chapter, the presentist needs to assume that only one past is compatible with how things are at present. But even with this assumption, since presentism confines reality to a single, durationless point, it cannot build into the description of how things are now any reference to motion, for this (assuming the truth of the static analysis) imports states at times other than the present. All that is available to the presentist is the position of objects, their various states, and the forces acting upon those objects. Is this enough to determine earlier and later positions? Consider the following two cases:

(i) The absolutist conception of space as an entity existing independently of its contents is correct. Consequently, there is such a thing as absolute motion (this being simply motion relative to space itself). Now consider a universe of objects in absolute, but not relative, unidirectional and non-accelerating motion, and on which no forces are acting to disturb their uniform motion. Can the position of those objects at any one time determine earlier and later positions? No, for the position of the objects at any one time gives one no information about the *direction* of motion, and there are no forces that determine that direction.

(ii) The relationist conception of space is correct, and all motion is relative to other contents of space. Now consider a (relatively simple) universe in which the forces on certain objects at a particular moment cancel each other out, so there is no resultant force on those objects in any one direction. Can the state of the objects at that moment determine their earlier and later positions? No, for the same reasons given under (i).

So, whether one is an absolutist or a relationist about space, there are situations in which present fact is simply not enough to determine whether objects *were* or *will be* in different positions from the ones they currently occupy. But on the static account of motion, whether something is now in motion depends precisely on what its past, present, and future locations are. So, in these situations, a combination of presentism and the static account of motion just leaves it indeterminate whether the objects in question are in motion or not. It seems, then, that it would be wise for the presentist to reject the static account of motion, and accept the dynamic account. And this, after all, is precisely what we would have expected.

Now, according to the version of the dynamic account which we were left with at the end of the last section, an object's being in motion is a primitive event, not further analysable in terms of objects, properties, and times. Now for these primitive events to exist, on the presentist reading, they must be capable of existing in the present. But events, being changes, are not instantaneous items: they take up time. So, at best, what exists in the present are *parts* of events. The idea of events having parts that are not themselves events, however, conflicts with the primitive status of events. To the question, what are these parts? the obvious answer seems to be: instantaneous states of an object. Presentism is therefore incompatible with the suggestion that events are primitive, unanalysable entities. And this is very bad news for the presentist, for that means it conflicts with the dynamic account of motion, and we have just given reasons why presentism should reject the static account.

Making use of these insights, we can now present our third and final version of the Arrow, as follows:

The Arrow, third version
   1. If motion is possible, then either the static or the dynamic account is the correct account of it.
   2. If presentism is true, the static account of motion is false.
   3. If presentism is true, the dynamic account of motion is false.
   Therefore:
   4. If presentism is true, motion is impossible.

We suggested at the end of the last chapter that if the universe were a B-universe, this would raise certain difficulties concerning our understanding of change. It seems we can now draw a similar conclusion about the A-universe. For, although the Arrow is specifically about motion, it is an easy matter to adapt it so that it concerns change in general. And if presentism is the only way for the A-theorist to avoid McTaggart's paradox, then, since the presentist has difficulty in accounting for change, it is not clear how we can accommodate change in an A-universe.

The problem is further compounded for that view of time, discussed in Chapter 2, which regards time as a series of changes. For if only the present can be real, time just is change, and change cannot occur in the present, does it not follow that time itself is unreal?

> **Questions**
>
> What is the smallest period of time possible? How far could something move in that period?
>
> Does a moving object move *in the present*?
>
> If only the present exists, and change must take time in order to happen, is change real?

CHAPTER 10

Interfering with History

Do not be afraid of the past. If people tell you that it is irrevocable, do not believe them.

Oscar Wilde, *De Profundis*

The Lost Days

In the September 1752 issue of *The Gentleman's Magazine*, a monthly periodical offering a diverting mixture of news, book reviews, recipes, advice, and poetry, there appeared a rather unusual letter. The correspondent reported an unsettling experience he had just had:

I went to bed last night, it was Wednesday *Sept. 2*, and the first thing I cast my eye upon this morning at the top of your paper, was *Thursday, Sept. 14*.

As the author explained, having until recently been in the West Indies, he had been unaware, until the fact had caught him out, that Parliament had at last consented to move from the Julian ('Old Style') to the Gregorian ('New Style') calendar. The correction required the removal of 11 days, so 3 September to 13 September, 1752 had been ruled out of existence. This had some amusing consequences, as the correspondent went on to explain:

I used to laugh at a man of my acquaintance for having a birth day but once in three years, because it fell on the *29th* of *February*: he laid me a considerable wager one of these nights, that I should lose a birth day some year or other as a punishment for my mockery. He was drunk when he made the proposal, but I little thought I should live to see it demanded. Sir, I am born the *13th* day of *September*.

And now the tone becomes rather aggrieved (though no doubt with tongue-in-cheek):

But must I confess the fatal truth to you. Sir, I have solicited the most amiable of her sex these five weeks: she seemed for a long time only to laugh at me, though my fortune is equal to her own: at last, Sir, she fixed the day, for the tenth of *September*, and gave me a bond of ten thousand pounds for the performance. I have consulted my lawyer; he is now at breakfast with me: and he says it will not do for next year, because the date 1752 is fixed to it: and so my ten thousand pounds are not worth ten pence. A fine affair, Sir, that a man must be cheated out of his wife by a parcel of *Mackmaticians* (sic) and *almanack makers*, before he has her: a new sort of divorce, truly.

Another correspondent, this time to Samuel Johnson's periodical *The Rambler* (for 26 March 1751) saw interesting possibilities in the change:

I think the new stile is a delightful thing; for my mamma says I shall go to court when I am sixteen, and if they can but contrive often to leap over eleven days together, the months of restraint will soon be at an end. It is strange, that with all the plots that have been laid against time, they could never kill it by act of parliament before. Dear Sir, if you have any vote or interest, get them but for once to destroy eleven months, and then I shall be as old as some married ladies . . . nothing surely could please me like a year of confusion, when I shall no longer fix this hour to my pen, and the next to my needle, or wait at home for the dancing-master one day, and the next for the musick-master; but run from ball to ball, and from drum to drum; and spend all my time without tasks, and without account, and go without telling whither, and come home without regard to prescribed hours, or family-rules.

I am, Sir
Your humble Servant,
PROPERANTIA

Not everyone took the change in such a light spirit. In many other European countries, the change had been made some 170 years earlier, on the recommendation of Pope Gregory, from whom the new calendar took its name. It would have been made in England at the same time, as Queen Elizabeth was in favour of it, had it not been opposed by the Archbishop of Canterbury, who was at pains to resist all papist influence. When the New Style was finally adopted, there was rioting in London and Bristol, just as there had been in Frankfurt in 1582, and there were less violent demonstrations elsewhere. Some people actually lost their lives in these upheavals. There seems to have been a general feeling amongst the demonstrators that some fundamental and unwelcome change had taken place. It was not just that a certain day was to be given the name of 14th, rather than 3rd, September, but that eleven days had really been cancelled. Would Saints' days now be celebrated *at the wrong time*? Had people been cheated out of wages rightly theirs? Had their lives been shortened? There was undoubtedly widespread confusion as to what the change really amounted to, whether it was no more than a correction to a convention system of dates, or genuine interference with history, indeed with time itself. If we are inclined to regard these incidents as confined to a less sophisticated age, we should remember that as recently as the second half of the twentieth century, a group of farmers in Midwest America, when faced with the introduction of Daylight Savings Time, objected that an extra hour of sunlight would burn the grass.

Well, we certainly cannot affect time by changing our calendar, or our clocks, but can we interfere with it by any other method?

The Alterability of the Past

There is a simple objection to the idea that we could have any effect on what has happened, namely that to affect the past would be to change it, and changing the past would involve a logical contradiction. Take some event, such as my gambling a preposterous amount of money in a game of roulette last night, and (of course) losing. Naturally, I regret my actions, and would like nothing more than to cancel that event, make it the case that it never happened. Unhinged by remorse, I persuade myself that I can indeed undo last night's disaster, and set about muttering the relevant incantations. Have I any hope of success? Suppose that I do indeed make it the case that I did not gamble last night. Since the fact that I did gamble and lose was the cause of my undoing that event, it appears to follow that it is both the case that I *did* and that I *did not* lose a large amount of money in roulette last night. This plainly being a contradiction, it follows that I cannot alter the past. But then if I cannot alter the past, I cannot affect it, for what is it to affect something if it is not to alter it? I affect the cushion I am sitting on by depressing it; the tea-pot affects the atmosphere of the kitchen by making it slightly more humid; the leaves falling outside are affecting the state of the house by blocking the gutters, causing them to overflow when it rains, and so making the walls damp. To affect, it seems, is to change. I cannot, on pain of contradiction, change the past, so I cannot affect it.

There is something worrying about this argument. If we apply it to the case of the future, and on the face of it we can, the result is a much less congenial conclusion. For can we change the future? To change it suggests that there is already a way in which the future is going to be, otherwise it is not there to change, so to speak. So let us say that it is now true that I shall step on the 10.05 train from Skipton to Carlisle tomorrow. I decide, at the last moment, not to go. Have I changed the future? If so, then it is the case now both that I will step on that train, and that I will not. I cannot avoid the contradiction by saying 'It is true now that I shall step onto the train, but, tomorrow morning, it will not be true that I step on the train,' for if there are truths about the future, then it is true now that I shall decide at the last moment not to step onto the train. I cannot, on pain of contradiction, change the future. But if to affect something is to change it, then I cannot affect the future either. That means that what will happen has nothing to do with my current decisions, and that I am therefore a helpless pawn in the hands of fate.

Some people have found that line of reasoning attractive, but it seems only prudent to look for ways of resisting it. One way is to reject the suggestion that there are truths about the future in the way that there are truths about the past. So, although it is now true that there was a thunderstorm over the Aire valley *last* week, it is neither true nor false on this view that there will be a thunderstorm over the Aire valley *next* week. There are no future facts, as there are past facts, and so nothing to make true, or false, assertions about the future. The future can be conceived of as a range of possibilities which are only resolved into one actuality when the moving present arrives there, rather like a zip fastener. That is one kind of A-universe, considered in Chapter 8, and it captures what many people think of as the asymmetry between the past and the future. On this view, the future is alterable in just one sense: it can be made determinate by our actions. It is not that there is already a truth about the future that we are at liberty to falsify: that would plainly be absurd. Rather, there is no truth as to how the future will turn out, but when it becomes present, what states of affairs are actualized will depend, in part, on our actions.

This is the view that we saw run into difficulties in Chapter 8. If the past is real, and this explains why we cannot affect it as we can affect the future, then there are past facts. But these past facts are not compatible with present facts. That was the gist of McTaggart's paradox. Now we suggested that there were two ways out of McTaggart's contradiction (perhaps there is a third, but what is it?): the B-theory and presentism. So let us see what consequences each of these views has for the alterability, of affectability, of the past and future.

In the B-universe, there is no passage of time, and no time that is uniquely and mind-independently present. All times, in consequence, are equally real. So if I make a statement about the past ('The bus was delayed this morning'), that statement has a determinate truth-value, because reality contains the earlier facts that make it true or false. The past (i.e. what OCCURS—see Chapter 8—earlier than your reading this) therefore cannot be altered, on pain of contradiction. But, equally, if I make a statement about the *future* ('The electrician will be here this afternoon') that statement already has a definite truth or falisty, because reality contains the later facts that make it true or false. The future (i.e. what OCCURS later than your reading this) therefore cannot be altered either, on pain of contradiction. This now leads us back to the problem we encountered earlier. If it is the unalterability of the past that explains why we cannot affect it, then, if the future is similarly unalterable, we cannot affect it either. The worry, then, is that the B-universe deprives us of our role as agents. However, we need not jump to this conclusion. It is important to distinguish, the B-universe proponent will insist, between two kinds of alteration. One is an alteration in *things,* as when a building is demolished. The other is an alteration in *facts,* where we take a fact to be of the kind expressed by 'The Public Library BE demolished at 11.30 a.m. on Wednesday, 21 November 2001'. We can alter things, but not facts. I can be responsible for the demolition of the Public Library, and so change the state of the building, but I cannot change the fact that the Library BE demolished at 11.30, etc. Now it is plausible to say that affecting the world means changing it only if the kind of change we are speaking of is the first kind, that is, change in things, not in facts. (It is worth pointing out that some kinds of affecting might involve *prevention* of change. So, for example, I affect the vase that has just been knocked accidentally from the mantelpiece by catching it before it smashes itself to pieces on the floor.)

So, following this line of thought, I can affect something if I bring it about that it has one property at one time and another property at another time, and this variation over time in properties is what constitutes change in the object. But this does not imply that I have the power to change the *fact* that the object has this property at this time and that property at that time. We can, then, affect future facts without altering them. Now although this is good news, does it not commit us to the view, which some may find contrary to both reason and experience, that we can affect the *past*? For if we can affect the future without altering it, the unalterability of the past is then no obstacle to our affecting it. Well, in distinguishing between affecting and altering, we may have left the door open to the idea of affecting the past, but we are certainly not committed to it. For affecting is a causal notion, and causation is one-way, or *logically asymmetric* (see Chapter 5), so that *a*'s being a cause of *b* is not compatible with *b*'s being a cause of *a*. It is also, arguably, *temporally* asymmetric (to be discussed in the next chapter), so that a cause always precedes its effect. Now the B-theorist can exploit the temporal asymmetry of causation as follows: to affect something is to be a cause of one of its features. But, causes being earlier than their effects, I can only affect how something is later, not how it is earlier. Therefore, although I may affect the future, I cannot affect the past.

So much for the B-theorist's view of affecting the past. What of presentism, the view that only what is present is real? Here it seems that the door has been left open, not only to the idea of the affectability of the past, but also to its *alterability*. Consider the question we raised in Chapter 8: what, if only the present is real, makes statements about the past and future true? It is hard to resist the conclusion that, assuming that concrete facts are needed to make such statements true, they are made true by present fact. Now, given that what is present is certainly within the range of what we can alter, does it not follow that, according to presentism, we can alter the past—by altering what makes true our statements about it? This idea is well captured in the following passage from George Orwell's *Nineteen Eighty-Four*, where the hero, Winston Smith, is being interrogated by the sinister O'Brien:

An oblong slip of newspaper had appeared between O'Brien's fingers. For perhaps five seconds it was within the angle of Winston's vision. It was a photograph, and there was no question of its identity. It was *the* photograph. It was another copy of the photograph of Jones, Aaronson, and Rutherford at the Party function in New York, which he had chanced upon eleven years ago and promptly destroyed. For only an instant it was before his eyes, then it was out of sight again. But he had seen it, unquestionably he had seen it! He made a desperate, agonizing effort to wrench the top half of his body free. It was impossible to move so much as a centimetre in any direction. For the moment he had even forgotten the dial. All he wanted was to hold the photograph in his fingers again, or at least to see it.

'It exists!' he cried.

'No', said O'Brien.

He stepped across the room. There was a memory hole in the opposite wall. O'Brien lifted the grating. Unseen, the frail slip of paper was whirling away on the current of warm air; it was vanishing in a flash of flame. O'Brien turned away from the wall.

'Ashes', he said, 'Not even identifiable ashes. Dust. It does not exist. It never existed.'

'But it did exist! It does exist! It exists in memory. I remember it. You remember it.'

'I do not remember it,' said O'Brien.

How can O'Brien say that the photograph *never existed?* The answer lies in his view of the past. The interrogation continues:

'There is a Party slogan dealing with the control of the past,' he said. 'Repeat it, if you please.'

' "Who controls the past controls the future: who controls the present controls the past," ' repeated Winston obediently.

' "Who controls the present controls the past," ' said O'Brien, nodding his head with slow approval. 'Is it your opinion, Winston, that the past has real existence?'

Again the feeling of helplessness descended upon Winston. His eyes flitted towards the dial. He not only did not know whether 'yes' or 'no' was the answer that would save him from pain; he did not even know which answer he believed to be the true one.

O'Brien smiled faintly. 'You are no metaphysician, Winston,' he said. 'Until this moment you had never considered what is meant by existence. I will put it more precisely. Does the past exist concretely, in space? Is there somewhere or other a place, a world of solid objects, where the past is still happening?'

'No.'

'Then where does the past exist, if at all?'

'In records. It is written down.'

'In records. And—?'

'In the mind. In human memories.'

'In memory. Very well, then. We, the Party, control all records, and we control all memories. Then we control the past, do we not?'

To put O'Brien's disturbing vision in terms of the two views of time we are considering: treating the past as having a reality that puts it on a par with the present, as the B-theory does, leads to the absurd conclusion that the past is somehow *still going on* somewhere, which implies that it is after all not past but *present*. Since this is plainly false, we are led to presentism, which denies all reality to the past except its existence in present causal traces, such as records and memories. That is, statements about the past are true only in virtue of facts about present evidence. Since evidence can be destroyed, replaced or otherwise tampered with, it follows that what it is now true to say of what happened in the past can similarly be altered. That is, history itself (that is, the events, not merely the records) can be interfered with. On O'Brien's version of presentism, the past is alterable.

But O'Brien's presentism is a very extreme and implausible kind of presentism. For one thing, it limits the range of present facts that determine the truth or otherwise of statements about the past to present *evidence*, i.e. traces that we would readily recognize as establishing, or suggesting, what happened in the past. So, because the burned newspaper photograph could not be recognized for what it is, it no longer counts as evidence, and so is no longer capable of making true any statement about the photograph's past existence. And the absence of such evidence, for O'Brien, means that it is now true that the photograph *never* existed. Finally, O'Brien supposes that evidence can be altered to suggest a quite different set of truths about the past. That is, the past can not only be obliterated: it can be created. But presentism is not committed to any of these disturbing consequences. First, the present facts that make true statements about the past do not have to be limited to what *we* would recognize as evidence. Why should our powers of observation or deduction be relevant? What matters is that the past leaves its traces, whether discoverable or not. And if statements about the past can be made true by facts we are not necessarily in a position to discover, then it may not be true that we can obliterate all traces of the past. If, for example, it turns out that only one past history is compatible with the present state of the universe, then nothing we can do will make any difference to what it is now true to say about the past. Secondly, even if past traces are obliterated, that does not necessarily falsify a once-true statement about the past. Suppose O'Brien has destroyed all traces of the photograph. It does not follow that there is sufficient evidence to establish that it did *not* exist. That is, the presentist can say that it is neither true *nor false* that the photograph existed. Thirdly, falsified evidence may bear the traces of its manufacture, and so may not be sufficient to establish different truths about the past.

A final comment on O'Brien: it is quite unfair to depict a view of the past as real as implying that the past is *still going on*. Treating the past as real is not the same thing as treating the past as another region of space (although it may be analogous in certain respects to a region of space). By definition, the past is *not* still going on, for it happened before the present. Only an implicit assumption that only what is present is real will yield the conclusion that the past can only be real if it is going on now, but such an assumption is quite illegitimate when considering the implications of, for example, the B-theory.

We conclude, then, that although some versions of presentism may represent the past as alterable, this is not an essential implication of all versions of the theory. There are reasons, whatever view of time one takes, to resist the view that the past can indeed be altered. But there is one hypothesis that seems to imply the alterability of the past, namely the hypothesis of time travel, so to that we now turn.

Dilemmas of the Time-Traveller

What is time travel? Anyone who has read a time-travel story has an intuitive conception of what it is, but when it comes to defining it, we face certain difficulties. Consider ordinary, spatial, travel: we travel about in space in virtue of occupying different places at different times. (Strictly, what makes this travel as opposed to mere motion is that it is, to a certain extent, under our control. But let us ignore that for the moment.) Now if we simply substitute time for space in this definition to give us a characterization of time travel, we end up with something that is either trivial or nonsensical: to travel in time is to occupy different times at different times. If this just means that, e.g. we occupy 8.15 a.m. at 8.15 a.m. and 4.30 p.m. at 4.30 p.m., then we are travelling in time constantly. But this is certainly not what we mean by time travel, which implies doing something different from what everyone else is doing. It makes no sense, however, to say that we can occupy 4.30 p.m. at 8.15 a.m. (within the same time zone), for this suggests that those two times are simultaneous, which by definition they are not.

It seems, to overcome the difficulty, that we need to distinguish between the time of the time-traveller and the time of the world through which he is travelling. So, following a suggestion of David Lewis's, let us talk of the *personal time* of the time-traveller, where this is understood as the set of changes going on in the time-traveller and in his immediate vicinity (whose bounds are defined by the sides of the time machine itself). Thus, the time-traveller's personal time just *is* the advance of the hands on his wristwatch, the beating of his heart, the imperceptible lengthening of his hair and nails, his changing thoughts, the burning of a nearby candle (supposing there to be such an anachronistic timepiece aboard the time machine), etc., etc. *External time,* by contrast, is time itself, registered by changes outside the time machine. So, despite the names, external and personal time are not two different times, or two different dimensions, for the changes that constitute personal time also take place in external time. Applying this distinction, let us imagine a journey taken by the time-traveller to 2101. The journey has taken 100 years in external time, but in terms of personal time, only 5 minutes (say) have elapsed. That is, the changes within the time machine occurring during the journey are those that normally register 5 minutes of external time. As far as the time-traveller is concerned, then, only 5 minutes have passed: that is what his wristwatch tells him, and how it appeared to him subjectively, and he himself has aged physiologically only by that amount. But, stepping outside his machine, he discovers that 100 years have passed away.

It is tempting, then, to define time travel as a discrepancy between personal and external time. But this is not really adequate. For, whether we are time-travellers or not, we all have our personal time, that is we all change by varying amounts in a way that registers time itself. Indeed, not only us sentient creatures, but all objects have a personal time. The clock on the mantelpiece has a personal time, the soufflé baking in the oven has a personal time, as does the aspidistra on the window sill, the fading photograph on the wall, the dripping tap in the kitchen. By varying the rates of these changes, I may perhaps be said to induce a discrepancy between the personal time of these objects and external time. To take a particularly dramatic illustration of the point, imagine that you volunteer to take part in an experiment in cryogenic preservation. You permit yourself to be frozen for 100 years, during which time your metabolic rate and heartbeat will be slowed down to almost zero. You will age, in effect, only a few days during that long period. At the end of it, you emerge not very much older, physiologically speaking, than you were at the beginning of the experiment. You have no memories of the freeze, since you were deeply unconscious for the whole of it. For you it will have passed in an instant. A short period of personal time has taken 100 years of external time—a significant discrepancy, and you will certainly have the sense of having travelled forward in time. And, according to the definition of time travel we are considering, you have indeed travelled forward in time. But have you? If there is some doubt in your mind, consider this rather simpler and more familiar case. My watch stopped 5 months ago and I still have not got around to taking it to the jeweller's to be repaired. It simply has not registered, in the way that it normally does, the passing of time. Has my watch become a time machine, then? Surely not.

To avoid this difficulty, it would be better to consider the history of the time-traveller and time machine to be a discontinuous one in external time. Thus, when the time machine leaves for 2101, it simply does not exist in the intervening times. If it did, then we would still be able to see the time machine standing exactly where it was when the time-traveller activated the controls, absolutely motionless and apparently unchanging within. But, as everybody knows, when a time machine leaves for another time it *disappears*. We can still make good use of the personal–external time distinction, however, as follows: the journey, although discontinuous in external time, is nevertheless continuous in personal time. That is, a clock aboard the machine does not suddenly take a great leap forward (or backwards) when the machine arrives at its destination: processes within the machine continue as normal.

We still have not dispelled the air of paradox, however. Think of the events taking place aboard the time machine: the traveller consulting time charts, the whirring machinery, the digits on the chronoscope (the instrument that tells one what year one would be in if one landed now) whizzing past. Where in time are these events? They cannot be *nowhere*, and yet we suggested just now that the journey itself has no location in external time. Perhaps the answer is that the journey takes place in another time-series altogether. We may imagine this time-series 'branching off' from our own and rejoining it at a later (or earlier) date (Figure 22).

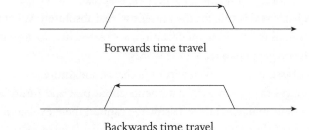

Forwards time travel

Backwards time travel

Fig. 22. Time travel and branching series

Whether or not the idea of two different time-series is coherent is a question we will leave until the next chapter.

Now we have a definition of time travel on the table, we can begin to explore the consequences of such travel. Let us say that backwards time travel, i.e. travel into the past, is possible. Does it not follow that the past is alterable? Or, to put the matter somewhat differently, if it is self-contradictory to assert that the past can be altered, does it not follow that time travel must be impossible? Let us pursue this line of thought.

The first point to make is that, for time travel to be possible, presentism cannot be the correct account of time. If neither the past nor the future is real, then it is impossible to travel to them. We could no more travel to 1789 or 2340 than to Samuel Butler's Erewhon. Travel into the past requires the reality of the past—and of the future. For, once one has arrived in the past, the present left behind becomes the future, and it would make no sense at all to suppose that the point of departure was not real (or no longer real). One might say: 'Ah, that future (or past) time might not be real *now*, but it will be real enough when you get to it.' But it would be very odd if reality depended on where one happened to be. If something is real, it is surely real *absolutely*, not merely real in relation to a particular location. (Consider the absurdity of 'Big Ben is real *in London*, but not real in Paris.' Is the temporal counterpart any less absurd?) Admittedly, the presentist does think that reality changes over time, but that is not the same thing as thinking that reality is relative to a time. What is real is still, for the presentist, real absolutely. What is now present is real without qualification, not real relative to the date at which you happen to read these words.

We will assume, then, the reality of both past and future, which in turn requires that we regard statements about past and future as having a determinate truth or falsity. We cannot, then, explain the alterability of the past in terms of the alterability of the present, on which the truth of such statements depend. But if statements about the past and future have a determinate truth or falsity, then we cannot change that truth or falsity, for to do so would result in contradiction, as we have already seen. It appears, then, that time travel requires for its possibility the absolute unalterability of the past (and future). And this conflicts with our intuitive understanding of what time travel makes possible. Suppose, for example, I discover that I left my umbrella on the train that is now a hundred miles away. Instead of feverishly jumping into a car and racing to the station at the end of the line, I calmly step into my time machine, set it for the time just before I alighted from the train, and this time make sure that I do not leave without my umbrella. What, after all, could possibly prevent me?

That this is not possible can be shown quite simply. The cause of my travelling back in time is a fact we will designate f (the discovery that I left my umbrella in the train, for example). The effect of my doing so is the negation of that fact, Not-f. But f and Not-f cannot both obtain. If this seems unconvincing, let us illustrate the problem by a more striking and appalling example. Disappointed in love, I wish myself dead. More than that, I wish that I had never lived. 'Let the day perish wherein I was born,' I say, along with the stricken Job. Given that I have a time machine, I am in a position to bring this about. So I travel back to some suitably distant moment before my conception, find a relevant relative (a grandparent will do if neither of my parents has yet been conceived) and, with malice aforethought, strike them dead. I thus bring it about that I was never conceived. But now this unsettling narrative must be exposed for the nonsense it is. If my action is successful, *who is it who prevents my conception?* It cannot be me, for it is now apparently true that I was never conceived, and so never grew up to step into a time machine to prevent my conception. I cannot, then, prevent my conception.

Take another example. Appalled by the disastrous and senseless loss of life during the First World War, I decide to travel back to 1914 and prevent the assassination of Archduke Ferdinand at Sarajevo. I spot the assassin in the crowd. I approach closer and closer and then . . . I trip over, and the shot is fired. I fail, and indeed I must fail, for if the assassination had not taken place then (let us suppose, perhaps not very plausibly) the First World War would not have happened, and so I would have had no reason to go back in time to prevent it. Dramatic though these two cases are, they are but two illustrations of the unassailable truth that I cannot change *any* past fact, however trivial.

This itself may seem to have further worrying implications. For if I cannot prevent my own conception, and cannot prevent the First World War, despite being present at the right time, does this not suggest that I am not, as a traveller into the past, a free agent? If this is an implication, then it extends to our ordinary, non-time-travelling situation. For I am not free to change the future, either. However, we can again appeal to our earlier distinction between *affecting* facts (i.e. being their causes) and *changing* facts. Although by travelling back in time I cannot change the past, I may yet affect it. Thus, my actions may—indeed, cannot fail to—have causal effects, effects which in part determine the character of the past. Suppose, as I approach the Sarajevo assassin, my tripping over causes me to fall against him, thus moving his arm just as the gun goes off. But, whereas (being a poor shot) he had been set to miss the Archduke, now, thanks to my interference, the bullet finds its mark, and history is set on its course of mass destruction. Of course, this is hardly a good example of a free action, but it does illustrate how I may affect the past without changing it. Time travel, then, does put the past within our causal reach, and so gives us control over it in just the same sense as we have control over the future.

But other paradoxes haunt the would-be time-traveller. We will end this section by considering two cases, each apparently allowed by time travel, but each containing an anomaly or contradiction. Here is the first case:

Tim is spending the summer holiday at his grandfather's house in rural Sussex. Bored one day, he wanders into his grandfather's library. On one of the more remote shelves, Tim discovers a dusty book with no title on its spine. Opening it, he sees it is a diary, written in a familiar hand. With a growing sense of wonder he realizes that one of the entries provides detailed instructions on how to build a time machine. Over the next few years, following the instructions to the last detail, Tim builds such a machine. It is finally completed, and he steps on board, and throws the switch. Instantly, he is transported back fifty years. Unfortunately, both the machine and book are destroyed in the process. Tim writes down everything he can remember in a diary. He cannot rebuild the machine, however, because it requires technology that is not yet available. Reconciled to getting back to the twenty-first century by the traditional method of doing nothing and letting time carry one back, he marries and has a daughter. The family move to a rambling mansion in rural Sussex. The diary is left to gather dust in the library. Years later, Tim's grandson, spending his summer holidays with his grandfather, discovers the diary.

The identity of Tim will be obvious, and this in itself is rather strange. But the question we are concerned with is this: where did the information on how to build a time machine come from? From the diary, of course, which itself was written by Tim. But where did he get the information from? From the very same diary! So the information has appeared from nowhere. At no stage has someone worked out for themselves how to build a time machine and passed on the information. The existence of this information is therefore utterly mysterious.

Here is the second case:

Peter and Jane, both 20 years old, are out for a walk one day in 1999 when suddenly a time machine appears in front of them. Out steps a strangely familiar character who tells Jane that he has an important mission for her. She must step into the machine and travel forward to the year 2019, taking with her a diary that the stranger hands to her. In that diary she must make a record of her trip. Obligingly, she does as she is asked and, on arrival, meets Peter, now aged 40. She tells Peter to travel back to 1999, taking with him the diary she now hands him, and recording his trip in it. On arrival in 1999, he meets two 20-year-olds called Peter and Jane, out for a walk, and he tells Jane that he has an important mission for her.

This raises several questions: How many trips are made in total? What happens to Peter and Jane? When they have finished travelling, what are their ages? But the really tricky question is: how many entries are there in the diary when Jane first steps into the machine? We imagine it blank. But this is the very same diary as the one Jane hands to the 40-year-old Peter, which by then contains her entry. And by the time Peter arrives back in 1999, it will contain his entry too. But then, if the diary already contained two entries when Jane was handed the diary, then it would contain three entries when she handed it to Peter, who would then add another one, so the diary would have contained four entries when it was first handed to Jane, and so on. If the problem is not immediately apparent, this is because we imagine an indefinite number of trips, but in fact there are just two: Jane's trip to 2019 and Peter's trip to 1999. So there ought to be a consistent answer to the question, how many entries are there in the diary when it is handed to Jane? Yet, as we have seen, there does not appear to be a consistent answer.

Causation in Reverse

Given that the time-traveller into the past cannot avoid interacting with a past time, a necessary implication of the possibility of time travel is the possibility of backwards causation. Let us say that the traveller lights a candle aboard the time machine just before departure. The candle is still burning on arrival, say, four centuries earlier. The cause of the candle burning in the seventeenth century therefore lies in the future: its being lit in the twenty-first century. 'Backwards causation' is the name given to the phenomenon, whether purely imaginary or not, where causes occur after their effects. Is backwards causation an incoherent notion? If so, then time travel is too. But, even if we reject time travel on other grounds, we can still ask about the intelligibility of backwards causation. Could something I do now affect the past?

One argument against the possibility of backwards causation has already been undermined: backwards causation is impossible, because it would involve affecting the past, and the past is unalterable. This objection is not a compelling one given our earlier distinction between affecting and altering. We can affect the past without altering it. But there is a related objection that is harder to dismiss, and it goes as follows. We can affect the future because it is not yet fixed, that is, there is as yet no fact of the matter as to what will happen. Our actions, in affecting the future, help to make it determinate. But the past is not like this. Unlike the future, it is entirely determinate. There is a fact of the matter as to what happened in the past. So, because our present actions cannot make the past any more determinate than it is already, they cannot affect the past. Causation cannot be understood without supposing this fundamental asymmetry between past and future. That is why causes are always earlier than their effects. Of course, this objection is not available to someone who thought of the universe as a B-universe. Nor is it available to the presentist. It is only available to one who thinks of the past as real but the future unreal—precisely the position that, we suggested, was vulnerable to McTaggart's paradox.

A rather different approach is to consider the circumstances in which we would actually be justified in thinking that we had witnessed a case of backwards causation. Let us say that I habitually wake up 5 minutes before my alarm clock goes off. Reflecting on this, I start to entertain the idea that it is the *later* going off of my alarm clock that causes my *earlier* waking up. How do we test this hypothesis? If there is just the one occasion where I wake up before the alarm goes off, we cannot test it, but then we would have no reason to entertain the hypothesis. In the case we are considering, the alarm clock going off is in general preceded by my waking up. So the hypothesis under consideration is that the alarm clock going off is *in general* the cause of my waking up 5 minutes earlier. I can try to falsify this hypothesis by waiting until I wake up in the morning, and then trying to *prevent* the alarm from going off. If I succeed, then I show that my waking up was not caused by the alarm, at least on this occasion. And if I accumulate enough such instances, I falsify the backwards causation hypothesis. But let us say I fail, and indeed fail on every single occasion that I prevent the alarm from going off. Perhaps I knock it off the table, or do not reach out in time, or the off switch is broken, etc., etc. Then there is a rival hypothesis to the hypothesis of backwards causation that also explains our failure: my waking up is (somehow) the cause of the alarm going off. This is, of course, a rather foolish example, but the general principle here can be applied to any case. We cannot have evidence which confirms a backwards causation hypothesis but which does not also confirm a rival forwards causation hypothesis. The two are incompatible if we assume that causation is *asymmetric*, i.e. is such that necessarily, if A caused B then B did not cause A. In these circumstances, we would never prefer the backwards causation hypothesis to the rival forwards causation hypothesis.

A final thought: if there were a large number of instances of events caused by later, rather than earlier, events, would the world not be rather more surprising than it is? That is, would we not expect to witness sudden happenings that would be completely unpredictable on the basis of the antecedent conditions and whose real explanation lay in the future?

Questions

What should Winston have replied to O'Brien's question whether the past had real existence?

Does the possibility of time travel into the *future* require the reality of the *past*?

How many entries were there in the diary given to Jane?

CHAPTER 11

Other Times and Spaces

'But do you really mean, sir,' said Peter, 'that there could be other worlds—all over the place, just round the corner—like that?'

'Nothing is more probable,' said the Professor.

C. S. Lewis, *The Lion, the Witch and the Wardrobe*

Probability and the Multiverse

You are presented with a large urn containing, you are told, 100 table-tennis balls. Removing one of these balls you discover, much to your surprise, that one of them has your name on it. Can you conclude anything about the rest of the balls in the urn? Precisely 100 hypotheses fit the rather limited data, ranging from the hypothesis that only one ball has your name on it (this being the one you happened to pick out) to the hypothesis that all the balls have your name on them. Clearly, you would be unwise to come to any fixed conclusion at this stage, for any of these hypotheses could be true. But are they equally likely? The probability of your picking out a ball with your name on it depends on the proportion of such balls in the urn. So, were there just one ball with your name on it, the probability of your picking it out first time would be 1/100, a rather small probability. In contrast, if all the balls had your name on them, the probability would be 100/100, i.e. 1, making it absolutely certain that your first choice would result in a ball with your name on it. This, surely, makes it more likely that all the balls have your name on than that only one does. To put it in more general terms, you would be wise to prefer a hypothesis that makes the observed result very likely to one that makes that result very unlikely. Of course, this judgement is only provisional. As you continue to take balls out of the urn, and observe whether or not they have your name on them, your preferences may change. The point is, however, that your first observation gives you some reason for supposing that the ball you drew out is not unique.

Consider another example. You are examining a page of printout from a computer whose function is to generate a completely random sequence of numbers. Your eye is caught by the first line of numbers: 314159265358979323846. They seem oddly familiar. After a moment, you realize that they are the first 21 digits in the expansion of π. Intrigued, you check the rest of the numbers on the page and discover that they all match the expansion of π. Now, you do not know whether this is the one and only page that the computer has produced, or whether it is one of millions of pages, the computer having been producing its numbers non-stop for years, and this page has been deliberately selected by someone for your attention. What are you going to assume? If this is indeed the only page of numbers the computer has produced, then it is the most remarkable and unlikely coincidence that it matches exactly the first part of the expansion of π. On the other hand, if this is just one of millions of pages (and perhaps taken from the printout of one of millions of computers, all generating random numbers simultaneously for years), then it becomes less improbable. So, given the general principle appealed to a moment ago, that we should choose the hypothesis that makes our observation more, rather than less, likely, we have reason to suppose, just on the basis of what we have before us, that this page is not unique—that it is one of many such pages. As before, our assumptions may change with more data.

Now consider a third case, this time not a fictional one. For life to evolve, certainly in anything like the form in which we are familiar with it, the universe had to have certain features. For example, there had, at some stage, to be carbon available in significant quantities. There also had to be water. The temperature of at least some parts of the universe had to be relatively stable, and within a certain narrow range (defined by the freezing and boiling points of water), requiring a source (or sources) of warmth that was both stable and remained neither too distant from nor too near to the emerging life-forms. There had to be a significant variety, both of atoms, and of ways in which atoms could combine to form molecules. Both atoms and molecules had to be relatively stable, and yet capable of undergoing reactions with other atoms and molecules to form novel molecules without requiring extraordinary conditions. These features in turn required more fundamental conditions concerning both the internal structure of the atom, forces between objects, and conditions obtaining in the early stages of the universe after the Big Bang (supposing the Big Bang to have actually occurred). Even a slight difference in any of the fundamental physical features of the universe, such as the forces that bind the components of atoms together, electromagnetic forces, the masses of particles, and the rate of expansion in the early universe, would have made it impossible for life to have evolved. Some of the details of this story are, it would be fair to say, still in dispute. Yet, even if only part of it is right, the existence of life depends on what has been called the *fine tuning* of the universe. That life exists is an indisputable fact. Yet, when we contemplate the huge variety of possible ways in which the universe could have been physically constituted, and the very narrow range within these possibilities that are compatible with life, that particular outcome—the emergence of life—seems almost unimaginably improbable. Are we content with this conclusion? Or do we, as with the cases of the urn and the page of numbers, look for hypotheses that make our observations less improbable?

The best-known hypothesis that transforms the probabilities is, of course, that of the existence of God. If the universe were the outcome, not of blind chance, but of divine design, then it is no longer a remarkable coincidence that the physical constitution of the universe lies in the very narrow band of possibilities that is compatible with life. *Of course* a benevolent God would have constituted the universe so that it was compatible with life. Given the existence and nature of God, the emergence of life in the universe ceases to be almost vanishingly improbable and becomes certain. Some people see the fine tuning of the universe to be a new argument (or, perhaps, a new variant of an old argument) for the existence of God. But there is another hypothesis that changes the probability of life, one that does not involve a creator, and which some cosmologists are taking seriously: the *multiverse* hypothesis.

According to the multiverse hypothesis, ours is just one of a number—perhaps a vast number—of universes, each of which exhibits different physical conditions. Given enough of these universes, a wide range of possible atomic, electromagnetic, and gravitational forces can be realized. Some universes have a Big Bang somewhere in their history, some do not. In some, the expansion of the universe after the Big Bang is very slow, and leads to a Big Crunch. In others, it is very rapid. In some, there are no stable atoms. Others are composed almost entirely of helium. Some may contain only two-dimensional spaces, others four-dimensional. Very possibly, others just consist of empty space and time. The more such universes there are, and the greater the range of physical constitutions realized, the less unlikely it becomes that one of them will contain just the right set of circumstances to permit life. In other words, postulating a multiverse is like postulating that the page of random numbers that just happens to match the expansion of π is just one of many such pages, produced by many machines, running over many years. As long as our universe is unique, the fact that it contains life is (the hypothesis of a creator aside) remarkable. But once we see it as one of billions of universes, each with a different physical make-up, the fact becomes less remarkable. Indeed, we may even be tempted to say that, given enough universes, it was inevitable that one should contain the conditions necessary for life.

It is not our present concern to examine the correctness of the reasoning behind the multiverse hypothesis. It is enough to show that it has some plausibility, and is at least a rival to other, theistic, explanations for why the universe is as it is. What we are concerned to do here is to examine what the hypothesis implies. As it is sometimes represented, the many universes are just mini-universes, part of a larger, all-encompassing Universe. But what would make these mini-universes part of a single universe? Not, certainly, their physical constitution, for it is an essential part of the multiverse hypothesis that the different universes have different constitutions. (It does not matter if some of them are closely matching, just as long as, collectively, the universes exhibit a wide range of possible constitutions.) So we seem to be left with the suggestion that the universes are all part of the same Universe by virtue of occupying (different parts of) *the same space,* and existing simultaneously. But this presents some problems. It is customary to think of the universe as governed by physical laws—laws of motion, of gravitation, of electromagnetic forces, etc.—by which I mean whatever in the world corresponds to, or makes true, statements of law, such as Newton's Laws of Motion. These laws, it is also customary to think, are *universal*: they apply in all places and at all times. It makes no sense, arguably, to think of them as being localized. But, if there are really such things as laws, then it is important for the many universes to exhibit different laws, otherwise the fine tuning problem just raises its head again: why was only this, life-permitting, set of laws realized? So, if different mini-universes, each exhibiting different laws, occupy the same place, then we would have to think of the laws of nature as being localized. But what confines them to their spatial boundaries? Could they spread? And what happens if one universe interacts with another? What laws, if any, would determine the result of their interaction? Perhaps talking of 'laws' as features of the world is inappropriate, however.

We can make true law *statements,* certainly, but what (on one account) makes these statements of law true could be the dispositional properties of things, and these *are* appropriately conceived of as localized in space. But one would also have to include relations between things, such as forces, and again the problem of interaction between different universes arises: what relations obtain between things from different universes? The universes may, of course, be too distant to interact, but there is still no reason to suppose that, at some point in the future, interaction will not occur. And what of space itself? Could it be three-dimensional in some places but four-dimensional in others?

These problems are neatly sidestepped by abandoning the supposition that the universes all occupy the same space. After all, it is not clear what is gained by conceiving of them of parts of the same Universe, especially if they are not allowed to interact. So instead of thinking of a universe as being defined by the laws that obtain there, we could define it in spatial terms: a universe is a collection of objects that are spatially related to each other, and are spatially related to nothing outside that collection (this second clause is needed to prevent my slippers, for example, constituting a universe by themselves). So, the multiverse is a collection of *distinct spaces,* each unrelated to the other. An object in one universe bears no distance relations to any object in any other universe. Or, to put it another way, there is no spatial path from one universe to any other.

It should be emphasized that this interpretation of the multiverse hypothesis is a highly contentious one, and there are ways of interpreting it that do not involve rejecting the uniqueness of space and time. Nevertheless, enough has been said here, I hope, for us to take the idea of other spaces seriously. Are there any other contexts where we might prepared to contemplate the idea of different universes? Let us take a brief look at one.

Branching Space

Between 1801 and 1803, Thomas Young, then Professor of Natural Philosophy (physics) at the Royal Institution, conducted a series of now famous experiments on the nature of light. In one of the best known, light from a single source was made to pass through two very narrow slits and then onto a screen. The two beams interacted to form an 'interference pattern' of dark and light lines on the screen that is characteristic of wave motions. Imagine two stones being thrown together at exactly the same instant into a pond, but a few feet apart. Each stone, as it hits the water, sends out a series of circular ripples, extending outwards on the surface of the pond. As the two sets of ripples meet, they cancel each other out in some places, and reinforce each other in others. This same pattern was produced in Young's experiment, thus providing compelling evidence that light travelled in waves. But light also behaves as a stream of particles. The 'corpuscular theory' of light had been proposed well before Young's experiment, but it received significant confirmation in the latter part of the nineteenth century and the early part of the twentieth. That light travels in packets, or quanta, of energy—*photons*—was the fundamental assumption of Einstein's (1905) explanation of the photoelectric effect. This in itself was extremely puzzling. How could anything behave *both* like a particle, which has a very well-defined location and a definite momentum, *and* like a wave, which is spread out in space, and to which the notion of momentum does not obviously apply?

But there was a further puzzle, discovered much later. The interference pattern produced by photons would persist even if only *one photon* was allowed to pass through either slit at any given time. That is, the pattern of distribution of photons over time is exactly the same as that obtained when many photons are passing through the slits simultaneously. A single photon on its own, one would naturally assume, has nothing else to interfere with, and can only go through one slit. So what is going on?

Here is one proposal (one of many, it should be pointed out): *all possible locations of the photon are realized.* As the photon approaches the slits, the universe branches. In some branches, the photon passes through one slit; in other branches, it passes through the other. Strictly, we should not talk of a single photon here, but rather many photons, each confined to a single branch. Indeed, every object is multiply duplicated, including the slits, the screen, and any device used to determine which slit the photon passes through. Then, when the photons in their separate branches have passed through the slits, the branches fuse, so that there is, once again, a single photon. But the behaviour of that photon will be determined by the behaviour of the many photons that existed before the fusion of the branches. All one can observe at any one time is a single photon because one's observation necessarily takes place in one and only one branch; and this is also true of all one's counterparts, each making their own observations in the other branches.

Let us suppose that this extraordinary proposal is true. What does it imply about space? If everything is duplicated, many times, then space itself must be duplicated, for otherwise different objects would be occupying the same space at the same time. So the branching universe is also a branching *space*, where the branches are spatially related to a space that existed prior to the branching, but are not spatially related to each other (Figure 23).

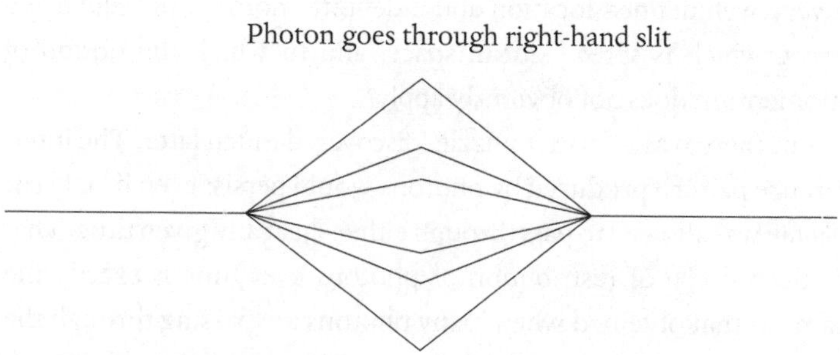

Photon goes through right-hand slit

Photon goes through left-hand slit

Fig. 23. Branching spaces

Again, this is just one of many explanations of the slit experiment, but it is at least an illustration of the way in which the idea of different spaces and times may be put to use. Our concern is not to examine the strength of the physical arguments in favour of the explanation, but to look at the conceptual difficulties it raises, and this we now do.

Objections and Consequences

The views of the world canvassed in the last two sections involve denying a feature Kant thought essential to space. As he put it:

[We] can represent to ourselves only one space; and if we speak of diverse spaces, we mean thereby only parts of one and the same unique space. (Kant 1787, 69)

This is tied up with Kant's view of space as a form of intuition, something that does not belong to things as they are in themselves, but a projection of the mind. It is not a projection we can cancel—we cannot but think of things as arranged in space—but it is a projection nonetheless. Space has no mind-independent reality. Now, in entertaining the multiverse hypothesis, or at least our particular gloss on it, we are treating space as mind-independent. So, we might think, we are not bound by Kant's insistence that there is only one space. But Kant presents the uniqueness of space as part of the motivation for his view of it as a form of intuition, not as a consequence. Is this just a stipulation, or a failure of imagination on Kant's part? Or is there some underlying reason why there can only be one space?

Here is one possibly relevant consideration. When we think of space, or of a collection of objects in a spatial arrangement, we think of ourselves in relation to it. We are, as it were, at the centre of our mental space. But, in imagining a group of distinct and unrelated spaces, we would have to imagine space in which we had no location, and this we cannot do. This is not a particularly compelling argument, however. There is a distinction between imagining and merely supposing. Imagination, or at least the kind of imagination that is being appealed to here, is closely related to perception. When we imagine objects in a spatial arrangement, we imagine perceiving (typically, seeing) them in that arrangement. The content of that perception will include information about how distant those objects are from us. So of course we are inevitably located in the space of our imagination. But not everything we can think of is imaginable in this way. I can contemplate the idea of four-dimensional space, as we did in Chapter 4, but I cannot imagine perceiving things *as* four-dimensional. I can conceive of a tree that I am not perceiving, but I cannot imagine *perceiving* a tree that I am not perceiving. We can, it seems, think of things without having at the same time to think of the relations we stand in to those things. Why, then, can I not think in objective terms about space, as a thing that does not necessarily have me at the centre of it?

Another objection to the idea of other spaces is that we could never have any evidence for their existence. Why not? Because, if there is no path from one space to another, there is no route by which one space could causally influence another: *causally* related things must also be *spatially* related things (that is, if anything like our ordinary conception of causality is correct). And evidence is a causal notion. We have genuine evidence for something when that thing causally affects us in certain ways. But this, too, is easily answered. Perhaps direct evidence is causal, but not all evidence is so. The fact that we cannot be affected by other spaces (or the occupants of those spaces) does not imply that we can have no reason to believe in their existence. The appeal to probability in the three cases discussed above (the urn, the random number generator, and the fine tuning problem) provides an example of non-causal reasoning. We can have evidence for other spaces—indeed, if the proponents of the multiverse are right, do have evidence for other spaces—that makes no appeal to the causal influence of those spaces on us. In any case, even if we restrict evidence to the causal effects of whatever it is that the effects are evidence of, the discussion of the slit experiment at least indicates the conceivability of our having such evidence for other spaces, for if those spaces are allowed to fuse at one point, they can exert a causal influence on each other.

Where we do have complete causal insulation of universes, as in our take on the multiverse, we have the suggestion of something more radical than the non-uniqueness of space, and that is the non-uniqueness of *time*. If there is no causal interaction between universes, then there is no way in which we can put the events occurring in those universes in any temporal order. According to the verificationist criterion of meaningfulness introduced in Chapter 1 (that a proposition we cannot even in principle verify as true or not is meaningless, or if not meaningless, then incapable of being true or false) that would be enough to prevent the truth of any given ordering of events from different universes in a single time-series. But that principle of meaningfulness is, we suggested, too restrictive, and we would be wise not to appeal to it here. But there is another consideration. We have in previous chapters noted the intimate connection between time and causality. Could it be that, where events cannot stand in any causal relation to each other, they cannot stand in any temporal relation either, because, ultimately, temporal relations are to be defined in terms of causal relations? We shall examine attempts at just such a definition in the final chapter, along with the difficulties such attempts face, but here we will note their implication: the multiverse is not just a collection of distinct spaces, it is also a collection of *distinct time-series*. Thus, just as there is no spatial path from one universe to another, the events in one universe are neither simultaneous with, nor earlier than, nor later than, any event in any other universe (see Figure 24). We can define a universe, in fact, as a collection of objects and events that are both spatially *and* temporally related to each other, and which are neither spatially nor temporally related to any object or event outside that collection.

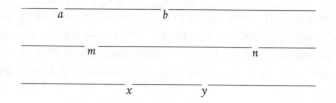

a is related to *b,* but not to *m, n, x,* or *y*

Fig. 24. Parallel time-series

As with space, one objection to this picture is that, in conceiving of a time-series, one necessarily locates oneself within that time-series. But it would be impossible for any individual to be located in more than one time-series, so, whenever we think of time, we must think of it as unique: each imagined event is imagined as temporally related to every other event. Now this objection may be rather more compelling than its spatial counterpart. Our ordinary experience of time is strongly suggestive of something that flows, or passes. That is (on one account of it), there is a unique and privileged moment, the present, that is constantly shifting from one event to another. Time, then, is fundamentally thought of in terms of an *A-series,* in the terminology we introduced in Chapter 8: a series of times or events running from the distant past to the distant future. The present is where, in time, we invariably are. The moves, then, are these: in thinking of a time-series, we necessarily think of an A-series, i.e. something which divides events into past, present, and future. So, in thinking of a time-series, we conceive of something part of which is (conceived of as) present. The present is where we are, so in thinking of a time-series, we necessarily locate ourselves in it. The analogous reasoning in the case of space is not nearly so plausible. There is no reason why we need to think of a space as something part of which is *here.* Now consider other time-series, series in which we are not located. Does it make sense to say, of any moment in one of those other series, that it is *present?* Arguably not. So, in thinking of time as essentially composed of an A-series, I must think of it as unique.

The person who thinks of the universe as a *B-universe*, that is, a universe in which there is no passage of time and in which all times are equally real (see Chapter 8), will have no such objection to the idea of other time-series. Indeed, they may even go so far as to say that the very fact that we can conceive of such unrelated time-series shows that we are not obliged to think of time in A-series terms. So, evidence for the multiverse is equally evidence for the B-universe.

Both lines of argument (that the notion of other times is incoherent because it conflicts with our conception of time as an A-series, and that our concept of time as an A-series is mistaken because the notion of other times is coherent) can be resisted, however. There need be no conflict between thinking of time as essentially an A-series and postulating the existence of other time-series. If time is, objectively speaking, an A-series, then its having a past, present, and future is completely independent of one's location and even existence. Times will continue to be present long after we have ceased to be. And although, when we are alive, our location always coincides with the present moment, that moment is not present *because* we are located at that moment, but rather the other way around: it is the presentness of some moment that constrains us to be at that moment and no other, given that we are somewhere in the time-series. So for a moment in some other time-series to be present, it is not necessary that I or anyone else be in that time-series. *If* we were in some other time-series, *then* we would be at the present moment in that time-series, but nothing requires us to be in any time-series at all. In other words, we can say:

> There exists another time-series, bearing no temporal relations to this one, and within that time-series some moment is present

without implying that that moment is also present in this series. (In fact, the implication is that the moment in question is *not* present in this time-series.)

But this is not the end of the matter. There were two components to the A-series view introduced in Chapter 8. One was the proposition that there is an A-series. The other was the proposition that the A-series is more fundamental than the B-series, in that it is the A-series facts that determine the B-series facts, and not the other way around. Let us call this second proposition the *reductionist thesis* of the A-series view. So, according to the reductionist thesis, it is because the ringing of the alarm clock and my waking up are both present that they are both *simultaneous*; and it is because the delivery of the milk is past and the ringing of the alarm clock present that the first *is earlier* than the second. As we might put it, statements about B-series relations between events are made true by underlying A-series facts. Now it is this thesis, rather than the simple statement that time consists of an A-series, that is put in doubt by the possibility of other time-series. Consider the time-series in Figure 24. We can coherently suppose both that *b* is present and that *x* is present, although they do not occupy the same time-series. But then, according to the reductionist thesis, those facts about *b* and *x* should make it the case that *b* is simultaneous with *x*, and, since they belong to different time-series, this is simply false. The falsity of the reductionist thesis does not entail that there is in fact no A-series after all, but it does make the idea that time is essentially an A-series much less attractive. For it was an advantage of the A-series view that it offered to explain the logical relationship between A-series and B-series facts, and the demise of the reductionist thesis would rob the A-series view of this explanatory feature.

There are at this stage two strategies the A-theorist could adopt. One is to come up with an analysis of B-series facts in A-series terms that is compatible with the possibility of other time-series. The other is to find some reason, other than that it conflicts with the reductionist thesis, for questioning the coherence of the idea of other time-series. Let consider the second strategy. Take another look at the sentence we used above:

There exists another time-series, bearing no temporal relations to this one, and within that time-series some moment is present.

For this to be intelligible, we need to read 'there exists' as having no temporal import. Thus, for example, when we say 'there exists a prime number greater than seven', we do not imply that there is a particular time at which this number exists, or that it exists *now* (but maybe not in the past). But when we are dealing with concrete objects, assertions of existence *do* have temporal import. At least, this is exactly what the defender of the A-series is likely to argue. One way in which we pictured the passage of time was that of events coming into reality when they became present. Prior to their being present, they were no part of reality. The future, in other words, is unreal. If this picture is correct, then there is a very intimate connection between assertions of existence on the one hand and time on the other, where this is construed as A-series time, i.e. a time composed of a past, present, and future. For the presentist, what is real just is what exists *now*. So the presentist's reading of the sentence above will be as follows:

There *now* exists another time-series, bearing no temporal relations to this one, and within that time-series some moment is present

and this is simply unintelligible, implying as it does *both* that the moment of question is present in our time-series, and that it is *not* present in this time-series. Allowing reality to the past but not the future presents a similar problem:

There exists, either now or in the past, another time-series, bearing no temporal relations to this one, and within that time-series some moment is present.

This implies, both that the moment in question is either present or past in our time-series and that it is neither present nor past in our time-series. So, granted that assertions of existence are *tensed*, i.e. they carry implications about our past, present, or future, the defender of the A-series is not obliged to contemplate the possibility of other time-series, as postulated in (one version of) the multiverse hypothesis.

However, there is a remaining difficulty for the A-theorist. Recall the idea, introduced in the previous section, of space branching into separate and unrelated spaces, which then, after a period of time, fuse. While the branches are distinct, they are causally isolated from each other, and their contents can only interact once the branches have fused. Applying once more the thought that causal unconnectibility implies temporal unconnectibility, we are led to the notion of branching or fusing *time*. Here, time-series which are separate at some points form a single series at others. Consider now the suggestion that our time-series is one that branches in the direction of the past, i.e. two separate time-series fused at some point in the past (Figure 25). From our current standpoint, both e and f are past. But, since they occurred in separate time-streams, they do not bear any temporal relationship to each other. Pointing out that existence assertions have temporal import does not conflict with this hypothesis of a branching past. For we can, without contradiction, say that there *were* in existence two separate time-series. Because they fused, both lie in our past. But now let us suppose that both e and f are exactly the same number of units past: let us say that they both occurred exactly 1,111 years, 11 months, 11 days, 11 hours, 11 minutes, and 11 seconds ago. Then, if A-series facts determine B-series facts, as the reductionist thesis asserts, they should be simultaneous. But e and f, occurring as they do in different time-series, are *not* simultaneous.

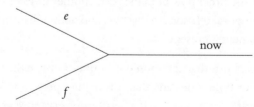

Fig. 25. The branching past

The moral of the story is that, if it is appropriate to interpret certain kinds of physical evidence, such as the fine tuning problem, and the two-slit experiment, in terms of non-standard structures for time, this puts some pressure on our view of the universe as an A-universe. And the difficulty does not so much concern the simple assertion that there exists an A-series, but the bolder assertion that naturally goes with it, namely that it is the A-series facts that determine the B-series facts, rather than the other way around.

Questions

If there are other worlds or universes, how likely is it that one of them is just like our own?

Could anything tell you that the universe had just branched into two? Could anything tell you that our universe had just fused with another one?

Near the end of one of his adventures *(Inferno),* in which he travelled to a parallel universe, Dr Who muses 'So not everything runs parallel. . . . An infinity of universes, ergo an infinity of choices. So free will is not an illusion.' Can you make sense of what he was thinking? Do you agree?

CHAPTER 12

The Arrows of Time

Einen Weiser seh ich stehen
Unverrückt vor meinem Blick;
Eine Straße muß ich gehen,
Die noch keiner ging zurück.

(I see a signpost standing immovable before my gaze;
I must go down a road by which no-one has come back.)

Wilhelm Müller, 'Der Wegweiser'

The Hidden Signpost

Could time run backwards? When we try to visualize this, we imagine something like a film being shown in reverse. But to imagine this is to imagine being *outside* the process we are observing. As an observer, we are not ourselves in the film (even if we are watching a film of ourselves, for the person we are watching on the screen is not, whilst been filmed, watching the film). So there is no conflict in thinking of the order in which we view the film as the reverse of the order in which the events I am watching were originally filmed. But, in imagining seeing *time* running backwards, we cannot absent ourselves in this way. We are not outside the time we are imagining. So we are not at liberty to see things in a different order from the order in which they are happening. Visualization, then, is no help. We have to think in somewhat more abstract terms when we entertain the idea of time going backwards. The very idea invites paradox. For if 'time runs backwards' means 'events are occurring in the opposite order from the order in which they happen', it is simply contradictory. And 'later events happen before the earlier ones' is no better. Even the time-traveller is not putting time into reverse. It is just that certain events in the time-traveller's life that would normally proceed in one direction are proceeding, for a time, in the reverse direction. What, then, if *all* processes went into reverse? Consider the way in which natural processes normally proceed: drops of rain fall from clouds, soaking the earth, rivers run downhill, the spring of a clockwork mechanism unwinds, heat moves from a hotter body to a cooler. Now imagine all these processes going in reverse: water travelling upwards from the ground to the sky, rivers running uphill, springs winding themselves up, heat moving from cooler bodies. That, surely, is not an incoherent description. But would this be a case of time running backwards? The very idea of all things going into reverse implies that time (or at least its direction) is independent of the processes within it, and indeed carries on normally while everything *else* goes into reverse.

In any case, although we may detect no contradiction in the idea of all natural processes going backwards, we can still ask whether this is really possible. Is there not some hidden constraint, some underlying directedness, within time itself, that makes processes within time go one way rather than another? This idea is captured by the talk of the *direction,* or *arrow,* of time. That time must have a direction seems required by our experience of it: our youth, once past, never returns. Yet, despite the all-pervasive sense we have that time is passing, and passing in one direction only, one of the greatest mysteries of time is what explains this direction. Indeed, it is no straightforward task to give a precise account of what it is that we are trying to explain. What does it actually *mean* to say that time has a direction? Once again, as with the passage of time, our understanding may be assisted by examining the contrast with space. Clearly, we are, within certain limits, free to go where we want to in space: we can go backwards, forwards, up and down, sideways in either direction. And any constraints there may be on our movement have nothing (we suppose) to do with the nature of space itself, but rather with the behaviour and influence of objects within space—we cannot, without creating special circumstances, escape the force of gravity, for example. (Of course, if space has an edge, or is otherwise finite, then that represents a restriction on possible movements, and in general other features of the space, such as whether it is curved, may influence objects, as we saw in Chapter 4. But this does not undermine the contrast we are trying to draw here between space and time.) In short, there may be directions *in* space, but no direction *of* space. In contrast, we are not similarly free to go where we want in time. I cannot, time travel aside, revisit yesterday. So we might expect to see this difference show up when we compare temporal and spatial relations.

Consider, then, the relation 'earlier than'. This relation orders things in time, and does so, we might suppose, in a way that reflects the direction of time. How does it reflect this? Well, as a first attempt at explanation, we could point to the logical properties of the relation: in particular, it is asymmetric: necessarily, if A is earlier than B, B is not earlier than A. So perhaps the direction of time just consists in the asymmetry of the earlier than relation. But this cannot be right. For the relation *to the north of* is similarly asymmetric, yet we do not think that, because of this, space has a direction. So something more has to be said. Consider another asymmetric spatial relation, 'to the left of'. This requires a point of reference for its application. The desk may be to the left of the window from where I am sitting, but if I were to turn around, it would be to the right. So 'to the left of' really needs to be considered as short for 'is, from such-and-such a position, to the left of'. 'To the north of' and 'above' similarly presuppose a point of reference, but, unlike with 'to the left of', this is a fixed reference point. 'To the north of' refers implicitly to the North Pole, and so means, in effect, 'nearer the North Pole than'. Similarly, 'above', in its usual application, means 'nearer the surface of the Earth than'. In contrast, 'earlier than' does not require a reference point in time for its application. If it is true at 11 o'clock that boarding the train this morning was earlier than my arriving at work, then it is equally true at 4 o'clock. Whether one event is earlier than another does not change over time. We can express the contrast by saying that 'earlier than' is *intrinsically asymmetric*, intrinsic because its holding between two items does not depend on the existence of any other item. But, although this contrast is significant, it still does not quite capture the distinction we are trying to draw between time and space. Consider another asymmetric spatial relation: *inside.* Whether one object is inside another does not depend, it seems, on any spatial reference point. Moreover, the relation can be used to order a group of suitably related objects, such as the collection of increasingly diminutive figures inside a Russian doll. However, the 'inside' relation is not a *pervasive* one: it cannot be used to order all objects in space, for many objects are not inside any other object. 'Earlier than', in contrast, *is* pervasive: by means of it we can order all events in time. So being in time at all requires that one stands in the 'earlier than' relation to other events or times. Now there are spatial relations that are both intrinsically asymmetric and pervasive, for example 'larger than'. However, it is not a *positional* relation: it does not tell one anything about where in space each object is vis-à-vis the other. 'Earlier than', in contrast, is positional: it tells one where events are in time vis-à-vis other events.

In saying that the 'earlier than' relation is intrinsically asymmetric, pervasive *and* positional, we undoubtedly capture something that distinguishes time from space. But, still, direction appears to point to something else. A series can be ordered without being directed. Consider a series of integers, ordered by the relation 'greater than':

114 is greater than 113 is greater than 112 is greater than 111 . . .

We have here an ordered series. But we are not obliged to read the series in one direction rather than another. We can start with the smallest and work up to the largest, or vice versa. Nothing in the numbers themselves picks out a preferred direction. There is no hidden signpost, as it were. But time is both ordered *and* directed, or so we suppose. Certainly the passage of time indicates a preferred direction in which events occur: the present moves *from* earlier events *to* later events, never the other way around. It is an interesting and important question whether the idea of a direction of time has to be tied to the idea of the passage of time, a question we shall come back to later in the chapter.

Three Arrows, and Why Things Fall Apart

We have made some progress in trying to capture what is distinctive about time, and it is likely that this is intimately related to the idea of the direction of time, but we have not yet arrived at an analysis of this elusive direction. So let us try a different approach. What draws our attention to time's arrow is the behaviour of processes in time. Some processes, it seems, are *temporally asymmetric*: some stages in the process characteristically come before others. These different processes, in effect, each constitute a different arrow of time. Here are the most significant:

⇒ the *thermodynamic arrow:* the direction from order to disorder;
⇒ the *psychological arrow:* the direction from perceptions of events to the memories of those events;
⇒ the *causal arrow:* the direction from cause to effect.

In each of these cases, the direction of the process coincides with the direction from earlier to later. Thus disorder tends to increase (in a sense to be explained shortly), memories always come after, never before, the perceptions of which they are the memories, and causes always precede their effects (or do they? we take up this issue later). This raises some tricky questions:

- *Why does each arrow coincide with the direction from earlier to later?*
- *Why do all three arrows point in the same direction? Why, for example, should the direction from experiences to memories coincide with the direction from order to disorder?*
- *Is one arrow more fundamental than the others?*
- *Can we actually define the direction of time in terms of one of the other arrows?*

Answering some of these questions helps us answer others. For example, providing an answer to the first also answers the second. For another example, suppose, in answer to the fourth question, that we can indeed define the direction of time in terms of one of the other arrows. That would then answer the first question in relation to that arrow. For if the direction of time is to be explained in terms of, e.g. the psychological arrow, then it follows as a logical consequence that perceptions precede memories. It would also be grounds for asserting that arrow to be more fundamental than the other two, and a test of that assertion would be whether, by taking that arrow to be more fundamental, we can explain why the three arrows coincide with each other.

We will now consider each arrow in turn, bearing these questions in mind.

Take the thermodynamic arrow first. According to a familiar and informal statement of the Second Law of Thermodynamics, heat tends to move from a hotter body to a cooler one, thus warming it up. This means that there is a tendency for heat to become ever more widely and evenly distributed. What has this got to do with order? Well, consider a cup of freshly poured tea steaming away on the table. This is a relatively ordered situation, because the energy present in it is concentrated in a relatively small area. There is energy in the form of heat in the small area occupied by the tea. There is also energy in the forces holding the cup together. Finally, there is potential energy in the cup as a result of its being some distance above the ground. But now a careless hand accidentally sweeps the cup off the table. The cup descends, dashing itself to pieces on the floor, and spilling the tea around the room. The potential energy of the cup was transformed into kinetic energy as it fell, and then into sound and heat as it smashed on the floor. The heat from the tea is now rapidly dispersing itself into the air. Thus energy is more evenly distributed. This little vignette, often used to illustrate the Second Law, is but one instance of a wider phenomenon, also instanced by stones dropping into ponds, causing ripples to spread out to the edges, or by sunlight warming the bricks of a house, and by the tendency of things—buildings, books, and bicycles—to fall apart. In these cases, energy is becoming more widely dispersed, and this is what is meant, in this context, by increasing disorder, and it is referred to as an increase in *entropy*.

A word or two about the definition of entropy and some other relevant terms. Take a particular item or set of items one happens, as a scientist, to be interested in. It might be a volume of gas, or of liquid, or a solid lump of matter, or an array of objects in interaction with each other. Let us designate this item or set of items by the abstract name of 'system'. Now suppose a definite amount of heat to be introduced into the system: the gas is briefly warmed by the sun, or the liquid by the application of a flame, etc. Then the *change of entropy* of the system as a result of this introduction of heat is given by dividing the amount of heat introduced by the temperature of the system prior to its introduction. Where heat is lost by the system, the entropy change is negative. Change in the entropy of a system, however, does not require the introduction or loss of heat from the system; it can instead result from the transfer of heat from one part of the system to another. Suppose we place a very hot object next to a very cold object, and consider the two objects together as a single system. Heat flows from the hot body to the colder, so both objects experience a change in their entropy: the hot body experiences a negative change, the cold body a positive one.

The overall heat within the system is unchanged. However, the positive entropy change in the cold body is *greater* than the negative entropy change in the hot body (since in the former case one is dividing the amount of heat by a lower temperature), so the two entropy changes do not cancel each other out. The entropy of the system as a whole goes up. This gives us a more formal statement of the Second Law, as follows:

Where any change occurs in an isolated system, entropy either stays constant or increases.

An 'isolated' system is one from which no energy can leave and into which no energy can be introduced. But what *is* entropy? One way of conceiving it is as follows: the entropy of a system is the degree to which the thermal energy (heat) in the system is so ordered that it is available for conversion into other forms of energy. The less available the energy is, the greater the entropy. So why is it that entropy tends to increase? Why do things fall apart?

One influential way of addressing this question is in terms of *probability*. Suppose you have, to continue with tea imagery, a large tea tray, and in one corner of this tray you have placed a number of sugar lumps. At the moment, they are neatly organized in two rows, one on top of the other, each row containing five lumps. You now begin to shake the tea tray vigorously and then set it down on the table. Assuming that the tray is flat and the table top not sloping, what are you likely to see? Not the same two neat rows of sugar lumps in one corner of the tray, that is for sure. Nor will they have gathered themselves neatly in the middle. Almost certainly, the lumps will have distributed themselves randomly around the tray. However many times you try the experiment, the result is likely to be the same. In other words, it is extremely *improbable* that the sugar should end up in an orderly arrangement, and concentrated in a small area. And this is because, whereas there are only a few such orderly arrangements, there are many more random or disorderly arrangements. So, given that the sugar is moving around randomly, the chances of its ending up in an orderly arrangement will be very small, reflecting the very small proportion of orderly arrangements amongst the total number of possible permutations. The reason energy tends to become more randomly distributed is the same: the permutations of random distributions outnumber the orderly permutations, so the probability is vastly in favour of a more random distribution.

But actually these probabilistic considerations do not really help us to see why disorder *increases* over time. A highly ordered state is indeed improbable, for the reasons given above, but it is equally improbable whether we are talking of an earlier time or a later time. Given an ordered state at a particular time, say the temperature distribution in my glass of gin and tonic (with ice and lemon) at 12 noon today, pure probability gives us no reason to suppose that the state of that system will be more disordered at *later* times than at *earlier* times. No more reason, in fact, than there is to suppose that things will be more disordered to the east of the gin and tonic than to the west. Of course, processes do proceed from earlier to later, and not vice versa, but we cannot just assume this without building in the very thing we are trying to explain, namely the direction of time.

Here is a quite different approach. Let us try to define the arrow of time in terms of the thermodynamic arrow. In other words, what makes it true to say that one state of the universe is before another state of the universe *just is* that the first is more ordered than the second. So, *by definition,* disorder tends to increase. This makes the thermodynamic arrow more fundamental than the others, because it, and it alone, defines the direction of time. More formally, what we might call the *thermodynamic analysis* of time order goes as follows:

The thermodynamic analysis of time order: event A is earlier than event B if and only if the universe is, when B occurs, in a higher state of entropy than it is when A occurs.

We now have two explanations of why disorder increases on the table. The first of them, as we saw, is not a complete explanation, but we do not complete it by combining it, as we might be tempted to do, with the second explanation. We must *not* say 'Time order just is the direction from order to disorder, so where there is no thermodynamic arrow, there is no arrow of time either. And we know why disorder tends to increase: disorder is just a more probable state.' The first part of this statement renders the second redundant. If we have defined time order in terms of the thermodynamic arrow, then we do not need to hunt around for some *other* explanation of why entropy increases over time. It would be like wondering why everyone describing themselves as 'married' on a census form appears, on inspection, to have a spouse. Worse, if we say it is merely *probable* (even if overwhelmingly probable) that entropy increases, then we leave the door open for an occasional decrease in entropy. But in *defining* temporal direction in terms of entropy, we do not leave this door open at all: it becomes a necessary (indeed, trivial) truth that entropy increases. So probabilistic explanations of the thermodynamic arrow, one is tempted to say, diminish its philosophical significance, even if they increase its physical plausibility. Since we are concerned with the philosophy of time, we will confine ourselves in the remainder of this section to an exploration of the thermodynamic analysis of time order.

One consequence of the thermodynamic analysis is that a universe in which things were always as disordered as they could be would exhibit no direction of time at all, because there would be no (significant) changes in entropy in such a universe. This is only a problem, however, if we think that time must exhibit a direction in order to exist at all (we return to this issue in the final section).

The analysis implies that the thermodynamic arrow is the most fundamental of the three. But the test of this will be whether we can explain, by appeal to the fundamental nature of the thermodynamic arrow, why the three arrows point the same way. There is some hope that the direction of the psychological arrow can be explained, along the following lines. The thermodynamic arrow points to a *global,* rather than merely local, change. That is, it is the overall entropy in the universe as a whole that is increasing. Locally, there can be decreases in entropy, as when, for example, a gas is compressed to the extent that it liquifies, or a boulder is rolled to the top of a mountain. But this local decrease in entropy is always at the expense of a global *increase* in entropy, because energy is required to induce local decreases in disorder, and this energy is ultimately dissipated in the form of heat. Now, storing information in the form of a memory is one such local increase in order which must be made up for by a global increase in entropy. So the direction from experiences to memories of those experiences will coincide with the direction from order to disorder. (We assume that it is just a matter of definition rather than accidental fact that the formation of memories constitutes a local increase in order.)

But what of the causal arrow? Can this be explained in thermodynamic terms? Well, suppose increases in entropy are always a result of causal processes. Certainly all our examples of increasing entropy were causal processes: the falling of a cup onto the floor, causing it to smash; the stone thrown into a pond, causing ripples to spread out to the edges; the warming of the bricks by the sun; the formation of memories, wear and tear causing things to fall apart. Causing something to happen involves the transfer of energy, and this transfer results in energy being more widely or randomly distributed. It is causation that brings with it the increase in entropy, so the causal arrow must point in the same direction as the thermodynamic arrow. Is that a satisfactory explanation? Well, we might think of possible cases where causation actually resulted in a *decrease* in entropy. Imagine, for example, God intervening at some point to bring about a more ordered universe. Now, if the thermodynamic analysis is correct, the direction of entropy increase is also the direction of time, so the case we have just imagined would have to be a case of backwards causation: God intervening to bring about a more ordered universe at an *earlier* time. But that, of course, is not how we imagined it: we imagined God bringing about a later state of greater order. So if what we imagined is a genuine possibility, the thermodynamic analysis cannot be correct. Now, defenders of the thermodynamic analysis will simply say that what we imagined is not a genuine possibility: not even God can bring about a decrease in entropy. So, unless we are willing to countenance backwards causation, the thermodynamic and causal arrows will always point in the same direction.

This illustrates, however, the implausibility of the thermodynamic analysis. As a statement of probability, the Second Law of Thermodynamics may well be true: it is overwhelmingly likely that entropy will increase. This leaves the door open for occasional, isolated instances of entropy decrease. But the thermodynamic analysis firmly shuts this door, and makes it a *necessary truth* that entropy will only increase. And this just seems too strong. How can we just rule out even the possibility of a once-in-a-blue-moon decrease in entropy?

Let us move on to the psychological arrow.

The Mind's Past

When the seventeenth-century English philosopher John Locke was looking for the thread that linked our past selves with our present self, the unifying feature that explained our persistence as persons through time, he alighted on memory. There is some plausibility in this, for total loss of memory robs us of our sense of who we are. Equally, loss of our ability to form new memories drastically diminishes our status as agents. In *The Man Who Mistook His Wife for a Hat* the neurologist Oliver Sacks describes the case of a patient he calls 'Jimmie', whom he first met in 1975. Jimmie was suffering from advanced Korsakov's syndrome, unable to remember anything for more than a few seconds. Moreover, although he could recall in detail events up to 1945, the 30 years afterwards were almost a complete blank. As a result, he thought of himself as he was in 1945: a 19-year-old, wondering whether to stay in the American navy or go to college. Sacks comments:

'He is as it were,' I wrote in my notes, 'isolated in a single moment of being, with a moat or lacuna of forgetting all around him. . . . he is a man without a past (or future), stuck in a constantly changing, meaningless moment.'

. . . 'I may venture to affirm,' Hume wrote, 'that we are nothing but a bundle or collection of different sensations, which succeed each other with an inconceivable rapidity, and are in a perpetual flux and movement.' In some sense [Jimmie] had been reduced to a 'Humean' being. (Sacks 1985, 28)

This apparently intimate connection between memory and the persisting self depends, however, on an even more intimate connection between memory and time. We remember only the past, never the future. And this is not just the trivial point that we would not *call* it 'memory' if we had experiences of the future; rather the past is revealed to us in a way in which the future is not. Why is this? Why does the psychological arrow point from earlier to later times? As with the thermodynamic arrow, let us see what happens if we try to define temporal precedence in terms of memory. Then we might get something like the following:

> *The psychological analysis of time order:* A is earlier than B if and only if A is the content of a memory at the same time that B is the content of a perception.

That would undoubtedly explain why memories succeed perceptions: they do so by definition. What happened earlier *just is* what someone remembers. But it is a rather surprising definition, for it makes temporal order dependent on minds. If there were no one around to have experiences of things, and subsequently memories of those experiences, there would be no earlier and later. This, for some people, would be enough to dismiss the idea that the psychological arrow is somehow fundamental. 'Surely,' they would argue, 'it is a purely accidental matter whether or not an event is actually experienced by someone? Moreover, experienced events may be caused by events that are not experienced, and it would be absurd to suggest that an experienced event was earlier than the memory of the event, but the causes of that event were not.'

There are two quite distinct issues here. One is whether two events have to be experienced by someone in order for one event to be earlier than the other. The answer to that, surely, is 'no'. The other issue is whether, in order for one event to be earlier than another, *some* event or group of events (but not necessarily that same pair) has to be perceived. Is it possible that temporal precedence does depend somehow on minds? It is far less obvious that the answer to *that* question has to be 'no'.

So let us try to define time's arrow in terms of the psychological arrow in a slightly more subtle way. First, let us introduce the relation of betweenness. We can order a series, for example, a group of people standing in a line, from information about which member is between which two other members. So, on being told that Betty is between Frank and Walter, that Doris is between Harold and Enid, that Harold is between Walter and Doris, and that Walter is between Betty and Harold, we can construct the line as follows:

Frank—Betty—Walter—Harold—Doris—Enid

In the words of the first section, this is an ordered, but not yet directed series. We can give it a direction by specifying that Enid is first in the queue, but betweenness alone will not bestow directedness on a series. Now, by analogy, we can talk of the relation of temporal betweenness, and order times, or events in time, by means of this relation. Consider now the following two facts:

(a) The letter arrived between the kettle boiling and Alf setting off to work.
(b) The kettle boiled before the letter arrived, and the letter arrived before Alf set off to work.

Since (b) implies (a), we would naturally assume that (b) is the more fundamental fact, and in general that facts about precedence determine facts about betweenness. But let us for a moment entertain the idea that (a) is the more fundamental fact, by virtue of its holding quite independently of the state of any mind, whereas (b) depends upon two mental states: an experience and a memory of that experience.

Pursuing this thought, suppose we take it for granted that, if e is an experience of some event, and m is the memory of that experience, then e is, by virtue of that psychological relationship, earlier than m. Now let us consider some quite different, unperceived, event—call it d—and let us suppose that e occurs between d and m. Now, although d is neither experienced nor remembered, we can deduce that it is earlier than e and m. Finally, consider another unperceived event, f, and suppose that m occurs between e and f. Then we can deduce that f is later than e and m. We can also therefore deduce that f is later than d. So, provided that some events are experiences, and that they are subsequently remembered, we can discover, of two unexperienced events, whether one precedes the other, given information simply about the relevant betweenness relations. This allows us to define temporal precedence as follows:

The modified psychological analysis of time order: A is earlier than B if and only if there exists an experience e and a memory of that experience m such that either

(i)   A is simultaneous with e and B is simultaneous with m, or

(ii) A is simultaneous with e and B is between e and m, or

(iii) B is simultaneous with m and A is between e and m, or

(iv) A and B are between e and m, and A is between e and B, or

(v) e is between A and B, and B is between e and m, or

(vi) A is between e and B, and B is between e and m, or

(vii) B is between A and e, but m is not between e and B, or

(viii) A is between m and B, but e is not between A and m.

This could hardly be described as an elegant and economical analysis, but it shows that it is possible for the fact that A precedes B to be mind-dependent, without this implying either that A and B have to be perceived, or that judgements about precedence are somehow subjective, that is merely a matter of opinion. For the analysis to work, we have to recognize that relations of temporal betweenness are entirely mind-independent. It is the mind, however, that gives time its arrow.

But is this analysis viable? Here is a problem. Suppose Monica has an experience, *a*, and subsequently a memory of that experience, *b*. According to the modified psychological analysis, this makes it the case that *a* precedes *b*. Now take two other events (whose nature will remain secret for a few more moments), *c* and *d*, which are so situated in time that *c* is between *a* and *b*, and *b* is between *c* and *d*. According to the psychological analysis, *c* must precede *d*. Now let us reveal the nature of *c* and *d*: one of them is an experience had by Norman, and the other is Norman's memory of that experience. But which is the experience and which the memory? Since we already know that *c* precedes *d*, it must be *c* that is the experience and *d* the memory if the psychological analysis is correct. But why should this be so? What is there to prevent *d* from being the experience and *c* the memory? In other words, why shouldn't the order according to Monica be the reverse of the order according to Norman? Norman's mental states are not determined by Monica's mental states: they are quite independent. And if time order is mind-dependent, there is nothing other than minds that determines the order of their mental states. There is nothing, then, to prevent different minds from imposing different orders on the world. But, if we allow this, then we have to say that time order is not just mind-dependent, but actually relative to individual minds. So our modified psychological analysis would become:

> *The relative psychological analysis of time order*: A is earlier than B *for a given person* if and only if that person has an experience *e* and a memory of that experience *m* such that either
> (i) A is simultaneous with *e* and B is simultaneous with *m*, or *etc.* (as before).

But this is simply implausible. Moreover, unless the causal and thermodynamic arrows are similarly relative to individual minds (and surely it makes no sense to suppose that the second of these, at least, is so relative), then there is no longer any correspondence between the psychological arrow and the other arrows.

It is time, then, to consider the causal arrow's claim to be the most fundamental.

The Seeds of Time

In Act I of Shakespeare's *Macbeth* the King of Scotland's generals, Macbeth and Banquo, returning from battle, encounter three weird sisters upon a storm-blasted heath. The witches greet Macbeth with the news that he has been made Thane of Cawdor, and will at some future time be king. Banquo, seeing his companion 'rapt withal' at this extraordinary prophecy, addresses the sisters:

> If you can look into the seeds of time,
> And say which grain will grow and which will not,
> Speak then to me, who neither beg nor fear
> Your favours nor your hate.

Why does Banquo talk of 'the seeds of time' rather than simply 'what is yet to come'? Well, perhaps for Shakespeare and his contemporaries the idea of the future somehow being available for observation by those with the appropriate gifts would have seemed nonsensical, as the future simply does not exist in any sense. The best one can do is to infer what will happen from its present causes, and it is these that are the seeds of time, the generators of history. Let us see, then, whether we can generate time order from the causal relations between things in time.

Can we define time order in terms of causal order? The simplest version of the causal analysis of time order is as follows:

The basic causal analysis of time order: A is before B if and only if A is among the causes of B.

It will be clear by now how the causal analysis could answer some at least of the other questions we posed about the arrows. The causal analysis makes the causal arrow the most fundamental of the three. Causes precede their effects because precedence has been defined in causal terms. And this explains why experiences precede memories of those experiences: the experiences are the causes of the memories and so, by the causal analysis of time order, must precede them. Can it explain why the causal and thermodynamic arrows (generally, but perhaps not invariably) point in the same direction? Well, the causal analysis cannot do this by itself, but if we grant that causation involves energy transfer, and such transfer tends to increase entropy, then both arrows will generally point in the same direction. And since the direction from earlier to later is defined as the direction from cause to effect, later states of the universe will tend to exhibit greater entropy than earlier states. So taking the causal arrow as the most fundamental would help to explain the temporal direction of the other two arrows. But it could do more than that. It also would explain why we cannot perceive the future. For perceiving something is a way of being causally affected by it. So, by the causal analysis, what one perceives must always occur earlier than one's perception of it. Perceiving the future would be an instance of backwards causation—an effect occurring earlier than its cause—and this the causal analysis rules out. The causal analysis of time order thus promises to be a very powerful tool. But it has some formidable objections to overcome.

Problems arise when we consider what to say of two events that are *not* causally connected. The causal analysis tells us that they cannot be related by the 'earlier than' relation. But this raises two issues. First, to put a rather obvious objection, it is surely possible for non-simultaneous events to be causally unconnected. Surely there could have been events in distant galaxies that have had no effect, and will have no effect, on us, and yet which lie in our past? Second, the basic causal analysis does not tell us whether events unconnected causally are simultaneous or completely unrelated temporally, but any adequate theory of time order would need to be able to distinguish between these quite different possibilities.

The first objection, that an event could be earlier than a causally unrelated event, could be met by replacing an actual causal connection with causal connectibility:

The modal causal analysis of time order: A is before B if and only if it is possible for A to be among the causes of B.

The term 'modal' here indicates that the notion of possibility is being invoked. What matters according to this analysis is not whether there is in fact a causal connection between two events, but whether there *could have been* a causal connection between them. If so, then the event that could have played the role of cause is the earlier. However, the shift from the basic to the modal causal analysis is fraught with danger. It is the appeal to possibility that is problematic, for there is more than one kind of possibility. The weakest kind is that of logical possibility, which we may define as anything that does not involve a contradiction. Thus it is logically possible that I could jump 20 feet in the air unaided, but not logically possible that I could prevent my own conception. More restricted notions of possibility would include reference to certain components of the actual situation. Given the law of gravitation and facts about my physical constitution, for example, it is not possible for me to jump 20 feet in the air. So what kind of possibility is appropriate for the causal analysis? If we choose the widest possible notion, namely logical possibility, then we face contradiction. Irrespective of how any two events, A and B, are related, it is always logically possible for A to be among the causes of B, and also logically possible for B to be among the causes of A (although both possibilities could not be realized together). It would therefore follow, from the modal causal analysis, that A is both earlier and later than B. But it is no good restricting the possibilities as follows: 'A is before B if and only if it is possible, *given their actual temporal relationship*, for A to be among the causes of B', for this would destroy the point of the causal analysis, in that it would make the temporal relation what determines the causal relation, and not the other way around. And if there was some more fundamental relation that determined whether A could be the cause of B, then one might as well refer directly to that relation, and leave out causation as redundant. So bringing in possibility achieves nothing.

Parallel Causes

Before introducing another, more promising, causal analysis, let us take a closer look at the second objection, that the basic causal analysis provides no means of distinguishing between simultaneous events and temporally unconnected events. Perhaps, in certain circumstances, we could use the causal account to distinguish between these two possibilities. Consider the two series of events in Figure 26. The arrows represent causal connections. So, in the first series, A causes B, which causes C, and so on. In the second series, α causes β, which causes γ, and so on. None of the members of the first series, however, is causally connected in any way to any member of the second series. According to the causal analysis, C precedes D, and γ precedes δ, but C neither precedes nor succeeds γ. But we are also able to infer, further, that C, for example, is not simultaneous with γ, for if it were, then B, being earlier than C, would be earlier than γ. But we have just said that B is causally unrelated to γ, and so cannot be earlier than it. According to the basic causal analysis, then, the two series are temporally unrelated.

However, although we have been able to use the basic causal theory to distinguish between simultaneity and temporal unrelatedness in this case, we have had to appeal to purely contingent features of the situation, namely that each series contains more than one member. What if there had been only two events, each causally

$$A \rightarrow B \rightarrow C \rightarrow D \rightarrow E \rightarrow F$$

$$\alpha \rightarrow \beta \rightarrow \gamma \rightarrow \delta \rightarrow \epsilon \rightarrow \xi$$

Fig. 26. Parallel causal series

unrelated to the other? If it is to be an adequate account, the causal account should tell us whether or not the two events are simultaneous.

Consider a quite different example (Figure 27), one in which the universe contained converging causal series. Here F and ξ jointly cause Z. What is the relationship between F and ξ? It cannot be one of simultaneity, for the reasons given above. If F and ξ were simultaneous, then E would be earlier than ξ, and that is ruled out by the causal analysis, which requires an earlier event to be a cause. So the result is that the structure above is, according to the causal analysis, one in which time itself branches in the direction of the past: two unrelated temporal series converge to form a single series. But this is not a desirable result. Arguably, our universe contains many such branching causal series: series initially unrelated to each other converge at some point to bring about a single course of events, which then leads to diverging series of causal chains which then have nothing to do with each other. We surely do not want to say that this implies that time itself exhibits a branching structure.

All the problems considered so far could be dealt with very simply indeed, by allowing the causal theorist to make explicit appeal to the relation of simultaneity, as in the following variant of the causal analysis:

The augmented causal analysis of time order: A is earlier than B if and only if A is simultaneous with one of the causes of B.

Since every event is simultaneous with itself, A's being a cause of B would satisfy the above requirement for being earlier than B. This

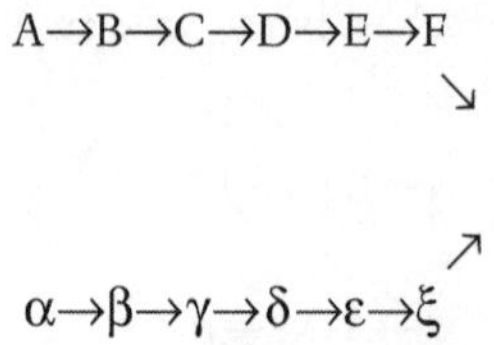

$$A \rightarrow B \rightarrow C \rightarrow D \rightarrow E \rightarrow F$$
$$\searrow$$
$$Z$$
$$\nearrow$$
$$\alpha \rightarrow \beta \rightarrow \gamma \rightarrow \delta \rightarrow \varepsilon \rightarrow \xi$$

Fig. 27. Converging causal series

new analysis allows (some) causally unrelated events to be temporally related. It also allows the causal theorist to specify under what conditions two events are temporally unrelated, namely when (i) they are not simultaneous with each other, *and* (ii) neither is the cause of the other. It therefore also allows us to distinguish between branching causal series and branching time. Of course, if 'simultaneous with' is defined as 'neither earlier nor later than', then we are simply smuggling in the very relationship we are trying to define into the analysis. But, as long as we allow for the possibility of multiple time-series or branching time, we will not want to identify simultaneity with the mere absence of the 'earlier than' relation.

Is Time Order Merely Local?

The introduction of simultaneity into the analysis raises a difficulty, however. We have already suggested, in our discussion of the psychological arrow, that temporal betweenness could be independent of temporal priority. In other words, it is possible for things to be temporally separated, without it being the case that some things occur earlier than others. Something else, not intrinsic to time, introduces the asymmetry. So, a proponent of the causal analysis could argue, a world where there were no causal relations could still be a world in which time existed, it is just that there would be no *arrow* of time. Only the existence of causal relations *somewhere* in the universe allows some times to be earlier than others. Now, the move from the basic to the augmented causal analysis was in part a response to the point that one event could be earlier than another even though no causal relation existed between them; all that is required is that they are appropriately related to items that are causally connected. But once this point is conceded, what is to rule out the possibility set out in Figure 28? Here, A is simultaneous with C, and B is simultaneous with D. C is a cause of B and D a cause of A. Since A is simultaneous with a cause of B it follows, from the

Fig. 28. Causal arrows running in opposite directions

causal analysis, that A is earlier than B. But B is also simultaneous with a cause of A, and so, again by the causal analysis, it also follows that *B is earlier than A*. This looks like a contradiction, but we cannot appeal to the causal analysis to rule out such a case, since it was the application of the causal analysis that enabled us to deduce the apparently inconsistent propositions that A is earlier than B and B is earlier than A. Nor can we say that the causal order of B and C is somehow constrained by the order of A and D, since the causal connection between A and D is quite independent of the connection between B and C. And we cannot appeal to any underlying asymmetry in time itself, forcing the causal arrows to point the same way, without undermining the causal analysis. Without causation, the analysis implies, there would *be* no earlier and later.

One quite tempting approach is simply to concede the result, but deny that it is at all contradictory. Only if we assume that temporal priority is a global relation is there a problem. Perhaps time order is simply *local* to causally connected series. We can allow, then, the existence of causally isolated series where the arrows point in opposite ways. In relation to *this* part of the universe, A is earlier than B; in relation to *that* part, it is B that is earlier than A. As long as they remain isolated, no anomalies will arise.

But what if these oppositely directed causal pairs do not remain isolated? Suppose, for example that the situation as in Figure 29 were to arise. Then, even granted that time order is simply local to a causal series, we still get a contradiction: A is both locally before and locally after B. We also get the contradiction that C is both earlier than, and simultaneous with, A. There are two possible ways out of this difficulty. The first, most radical, way is simply to *give up the notion of simultaneity altogether.* Without simultaneity, the problem cannot arise. We could, then, insist that without causation there is no time at all, not even undirected time. For there to be temporal relations of any kind between events, there must be causal connections. So, take a rather simple universe in which there are only a very limited number of events, and suppose that when we have mapped out all the causal connections we end up with the picture shown in Figure 30. Then we can say that both A and B are earlier than D and F, that D is earlier than F, and that C is earlier than E. Neither C nor E, however, stand in any temporal relationship to A, B, D,or F. What of the relationship between A and B? Well, although both are earlier than D, they are not themselves causally connected, and so neither is earlier than the other. In giving up simultaneity, we have moved back from the augmented to the basic causal analysis.

The less radical proposal involves appealing to certain facts about causation, and in particular the following:

The Betweenness Rule: If B is causally between A and C (e.g. an effect of A and a cause of C), then B is also temporally between A and C.

Fig. 29. A causal series changing direction

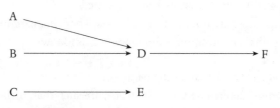

Fig. 30. A causal map

Appealing to this rule allows us to exclude cases where a series contains two causal arrows pointing in opposite directions. For example, the case we considered either, where A is simultaneous with C, C causes B and B causes A, violates the Betweenness Rule. This might at first sight seem a suspicious move. After all, surely we are trying to explain the properties of time in terms of the properties of causation, and not the other way around? But, as we have already indicated, there is no inconsistency in saying both that temporal betweenness is independent of causality, and that temporal priority is not. The Betweenness Rule does not, however, outlaw cases of different and isolated causal series where the causal arrows point in opposite directions. Unless we can find some means of doing so without undermining the causal analysis, it looks as if we have to accept that a consequence of that analysis is that time order is merely local.

Are Causes Simultaneous with Their Effects?

The causal analysis has not yet passed all the hurdles it needs to in order to be acceptable, however. It is a fundamental requirement of any causal analysis that causes precede their effects. But in some cases, cause and effect seem to be *simultaneous*: a crown resting on a cushion and causing an indentation in it, for example, or the passage of an electric current through a wire causing a magnetic field around the wire, or the engine of a train causing the carriages to move at the same time. If these genuinely are cases of simultaneous causation, then the causal analysis is doomed.

Perhaps these are not just isolated cases. There is an argument to the effect that *all* causation is simultaneous, as follows:

The simultaneity of causation argument
   (1) Causes determine their effects, i.e. are such that, if the cause occurs it is impossible for the effect not to occur.

(2) If there were a temporal gap between cause and effect, then causes would not determine their effects, for something could intervene in the gap and prevent the effect from occurring.

Therefore:

(3) There is no gap between causes and effect.

Therefore:

(4) Causes are simultaneous with their effects.

The conclusion of the argument is worrying, for it seems to be incompatible with the notion of temporally extended causal chains, and hence with the idea that earlier events are causally relevant to later ones. However, despite the appearance of plausibility, the argument fails. We could attack the first premiss, and insist that causation is in fact indeterministic, but the major defect is the move from the first conclusion to the second. If the effect occurs *immediately* after the effect, there is no gap.

In at least some cases, the suggestion that cause and effect are simultaneous leads to contradiction. For example, a spoon placed in a cup of coffee will cool the coffee down, but in doing so must, by the Second Law, itself heat up at the same time. Suppose the coffee's cooling down at time t' is caused by the spoon's being placed in the coffee at t. An essential part of the cause is the temperature of the spoon at t. Suppose, then, that the spoon is 20 °C at t. It must therefore be more than 20 °C at t', because in cooling the coffee it will have warmed up. If $t = t'$, therefore (i.e. if causation is simultaneous in this case) we get the contradiction that at t the spoon is both 20 °C and more than 20 °C. Another example is provided by colliding billiard balls. A red ball, moving with some speed across the billiard table, crashes into the stationary black ball at t, causing it to move at t'. The momentum of the red ball at t we will designate by m. It is this momentum that is an essential part of the cause of the black ball's acquiring a non-zero momentum. But the red ball must necessarily lose momentum in causing the black ball to gain it. So the momentum of the red ball at t' is less than m. The consequence of identifying t and t', i.e. of assuming simultaneous causation, is the contradiction that the momentum of the red ball is both m and less than m at the same time. The general principle in play here is that the state of an object must necessarily change when that object causes a change in something else.

What of the plausible cases of apparently simultaneous causation we presented above? The causal theorist will, of course, resist the suggestion that these have to be viewed as cases of genuinely simultaneous causation. It cannot be denied that we are presented with two states, or processes, that are simultaneous. The crown's presence on the cushion is simultaneous with the depression in the cushion. The motion of the engine is simultaneous with the movement of the carriage. But simultaneous processes, even simultaneous processes between which there is a causal connection, do not imply simultaneous causation. These processes have distinct parts: the movement of the engine at 10 o'clock is a distinct state from its movement at 1 minute past 10 (or even a split second past 10). And the presence of the crown at breakfast time is a distinct state from its presence at teatime, even if it has not changed at any time. These distinct states have distinct effects: it is the movement of the engine at 10, *not* at 1 minute past, that takes the engine past the signal box. It is the presence of the crown at breakfast-time, *not* its presence at teatime, that causes me to see the crown at breakfast-time. So we can resist the temptation to describe these cases as genuine cases of simultaneous causation by reflecting on the following redescription of them: the movement of the engine at a time, t, is the cause of the movement of the carriage shortly *after* t; the presence of the crown at t is the cause of the depression in the cushion shortly after t. To show that this is indeed the correct account of what is going on, bring the engine to a sudden stop. Does the carriage stop moving *immediately*? No, inertia carries it on a short distance, slightly compressing the engine. Lift the crown off the cushion. Does the depression disappear immediately? No, the cushion only gradually starts to reinflate.

A Sense of Direction in a Directionless World

The causal analysis, it seems, has been able to resist some of the objections thrown at it. But there is a question we left hanging at the end of the first section of this chapter, and so far we have not answered it. An ordered series, we said, is not the same thing as a directed series. We can order the series of integers, but we are not obliged to think of that series as running in a certain direction— from smallest to largest, say, rather than from largest to smallest. Now the causal theory of time order is just that: a theory of *order*. It gives an account of what makes an event earlier than another. But, we are inclined to suppose, the direction of time is more than that: it is the direction *from* earlier *to* later. So what makes this the preferred direction? We will end this chapter by looking at two quite different responses to this question.

As we saw in Chapter 8, there is more than one way in which we can order events in time. Events ordered by the *earlier than* relation constitutes a B-series, recall. But events may also be ordered as an A-series, that is, in terms of their pastness, presentness or futurity. A-series positions, we observed, *change*, so that what was once future becomes present and then ever more past. And this is the basis of the passage of time. So perhaps the missing element in our discussion of time's arrow is the passage of time itself. It is the present moving from earlier to later events, we could argue, that gives time its preferred direction.

Setting aside for a moment the difficulties, discussed in Chapters 8 and 9, that the idea of the A-series raises, we should ask whether the flow of time really does help us understand time's arrow. The idea is intuitively appealing: we imagine the present moving along the dimension of time in a particular direction. But what makes it the case that the present is moving towards the *future*, rather than towards the *past*? Of course, there is a trivial answer to this question, and that is that the future is just defined as that towards which the present is travelling. So let us rephrase the question. Suppose

ABCD represents a series of events in the order in which they occur. What entitles us to say that the present is moving from A to D, rather than from D to A? We can only answer this question in temporal terms: in order of time, A comes first. That is, A is, quite simply, *earlier* than D. But if this is the explanation, then it is hard to see what the A-series explains, as far as direction is concerned, that the B-series does not. It is true that, if the A-series exists, then events only stand in B-series relations by virtue of their A-series positions, so that A is earlier than D only by virtue of the fact that, e.g. A is past and D is present. But that dependence of the B-series on the A-series would not illuminate the issue of *direction*, because we cannot describe the direction in which time flows without appealing to facts about order, and order can be as well expressed in B-series terms as A-series terms.

Let me approach the issue in a different way. If events truly have A-series positions, then those positions change. Thus, to give the full picture of the passage of time, we have to describe, not just the A-series positions of events at any given moment, but also their A-series positions at other moments:

(i)   A is present, B is future, C is future
(ii) A is past, B is present, C is future
(iii) A is past, B is past, C is present

So, the answer to the question why the present moves from A to C, and not vice versa, is that the present moves from (i*) to (iii*):

(i*)   (i) is present, (ii) is future, (iii) is future
(ii*) (i) is past, (ii) is present, (iii) is future
(iii*) (i) is past, (ii) is past, (iii) is present

But now we have to ask why the present moves from (i*) to (iii*) and not vice versa. And so it goes on. At no stage do we get a full answer to our question. Clearly, it would be better not to embark on this fruitless quest in the first place, and say simply that A becomes present first because A is earlier than B and C.

What these reflections lead us to is the following radical thought: perhaps what we have been pursuing in our discussion of this difficult topic will turn out not to exist. We have been assuming that there is a difference between a merely ordered series and a *directed* one, and it is directedness that we have found so elusive. *But what if there is no such difference?* What if, in other words, the direction of time is nothing more than the asymmetry of the 'earlier than' relation? Well, the asymmetry of that relation does not in itself explain why causes occur before their effects, why disorder increases, and why memories only form themselves after the relevant experience. But there may be a more fundamental fact underlying all of these, namely the logical asymmetry of the *causal* relation: if *a* was the cause of *b*, *b* cannot have been the cause of *a*. If the 'earlier than' relation can be reduced to the more fundamental causal relation, then we not only explain why causes occur before their effects (they do so by definition), we also, it may plausibly be argued, explain why disorder increases and why memories follow experiences: by appeal to the fact that these processes are causal processes. So the direction of time is just the direction of causation. But, it will be objected, although we may explain why the direction from earlier to later is also the direction from cause to effect, we have not explained why the direction from earlier to later, and not from later to earlier, is *the* direction of time. To which it can be replied: we do not have to say that there is a preferred direction. To go on insisting that time must have a direction in this sense is rather like saying that all hills really go *up*hill, never *down*hill.

This spatial analogy will perhaps help us find the last piece of the jigsaw. Although hills themselves are not more uphill than downhill, our *experience* of them may be. Thus, we can experience a hill as going upwards rather than downwards, or vice versa: it just depends where we start from. So, perhaps, it is with time: we simply experience it as going one way. But, if there is no direction of time over and above time order, is our experience of apparent direction not precisely the problem that needs to be explained? We cannot rest content with 'There is no direction: we just experience time as if there were'; we have to explain our *sense* of direction in a directionless world. This is, essentially, the difficulty with which the hero of H. G. Wells's *The Time Machine* is grappling with in conversation with his incredulous friends:

'Clearly,' the Time Traveller proceeded, 'any real body must have extension in *four* directions: it must have Length, Breadth, Thickness, and—Duration. But through a natural infirmity of the flesh . . . we incline to overlook this fact. There are really four dimensions, three which we call the three planes of Space, and a fourth, Time. There is, however, a tendency to draw an unreal distinction between the former three dimensions and the latter, because it happens that our consciousness moves intermittently in one direction along the latter from the beginning to the end of our lives.'

'That', said a very young man, making spasmodic efforts to relight his cigar over the lamp; 'that . . . very clear indeed.'

'Now, it is very remarkable that this is so extensively over-looked,' continued the Time Traveller, which a slight accession of cheerfulness. 'Really this is what is meant by the Fourth Dimension, though some people who talk about the Fourth Dimension do not know they mean it. It is only another way of looking at Time. *There is no difference between Time and any of the three dimensions of Space except that our consciousness moves along it.*'

Evidently, our Time Traveller is a B-theorist: he has no use for the passage of time. (Although, to be fair to B-theorists, many of them would be quite unhappy to hear time described simply as a fourth spatial dimension. The differences between time and space do not all disappear with the demise of the A-series.) What then is his explanation of our sense of the passage and direction of time? It is that 'our consciousness moves along' (in one direction, note) the dimension of time. But if our consciousness really is moving along this dimension, and not just apparently so, then the very thing that the Time Traveller is attempting to banish has been reintroduced. For how are we to describe this movement of our consciousness except in temporal terms? Our consciousness *was* at that point of time, and *is now* at this point. If such movement is permitted, then there really is such a thing as the passage of time, and the difference between time and space, so passionately denied by the Time Traveller, is restored.

Perhaps, however, the talk of movement in this context is only meant metaphorically (although it is unlikely that the Time Traveller's friends would have so taken it). Perhaps all we have are different states of consciousness at different times. But how could this give rise to the impression of direction (and passage)? Think again of the experiential asymmetry between earlier and later times: we remember the past, perceive the very recent past, but have no experiences or other psychological traces of the future. We can explain this asymmetry by appeal to the asymmetry of causation. Since the direction from earlier to later just is the direction from cause to effect, it follows that backwards causation is impossible. But experiencing later events would be an instance of backwards causation, so such experience is impossible.

Finally, we may note one more useful feature of the causal analysis of time order: if it is true, it solves the mystery of why time has, indeed must have, only one dimension. The causal series is itself necessarily one-dimensional. There is only one way in which events can vary in their causal distance from other events. If time order just is causal order, then time too can have only one dimension.

Questions

Could time run backwards?

If time is all in the mind, why does it seem to have a direction?

Are there any cases where a cause is simultaneous with its effect?

Concluding Thoughts

'But,' said Tom, 'suppose someone really had stepped out of one Time into another—just like that—then that would be proof.'

'*Proof!*' cried Uncle Alan and for a moment Tom thought he was going to be angry again, but he controlled himself. 'I have been able to explain very little to you, Tom, if I have not even conveyed to you that proof—in matters of Time theory—*Proof* . . . !' Apparently, about Time, as about some master-criminal, you could prove nothing.

Philippa Pearce, *Tom's Midnight Garden*

What are space and time? Are they real, or do they exist only in the mind? And if they exist without the mind, are they objects in their own right? Or are they collections of relations between things and events? What are their features, and what explains why they have these features? Could they have had different ones? Are they, for example, infinite, or only finite? If finite, do they have boundaries? Are they infinitely divisible, or are they composed of 'atoms'? How does time differ from space? Does it really pass? Is the future real? And what accounts for time's direction?

These are some of the questions we have tried to address in the preceding pages, and we have done so, for the most part, through a study of the difficulties and paradoxes that our ordinary views of time and space throw up. This has really just been the start of an investigation, rather than an exhaustive enquiry, and I do not propose to offer a set of definite answers to the questions just posed. Instead, this last section of the book summarizes the preceding discussion, attempts to draw some of the threads together, and raises some further questions, questions to do with the human significance of our philosophical views.

Space and time have an unrivalled capacity to generate paradox. It is hardly surprising, in the light of this, that many of the philosophers who have written about them have concluded that they are unreal. Parmenides, Zeno (arguably), Kant, and F. H. Bradley denied the reality both of space and of time. Even Aristotle, who does not dispute their existence, acknowledges that they raise so many difficulties that it is quite reasonable to suppose that they do not exist. St Augustine concludes his long and searching exploration of time with the judgement that time is in the mind. McTaggart argued that there was a contradiction in the very notion of time. The consequences of denying the existence of space and time, at least outside the mind, are, however, significant. So much of our conception of the world is bound up with its apparent spatial and temporal features that to deny the reality of those features would imply that in investigating the nature of the world we are simply investigating the contents of our own minds. Kant embraces this consequence, arguing that only this explains how we come to have knowledge that is both *a priori* (i.e. such that recognition of its truth does not depend on particular experiences) and *synthetic* (i.e. not just a matter of definition).

One of the lessons of our inquiry is that the reality of space and time need not be an all-or-nothing matter. There is a considerable variety of features we ordinarily ascribe to space and time, and it is always possible to hold that one feature is unreal but another not (provided the features are not logically related). For example, we may believe in the objectivity of spatial and temporal relations without also believing in the objectivity of their *metric*. It may be true, independently of any mind, that event C occurred after event B, which occurred after event A, without there being any fact of the matter as to whether or not the interval between A and B was equal to that between B and C. According to *conventionalism over metric*, it is only according to a particular system of measurement, and not absolutely, that one interval is as long as another. Whether this system is the correct one is, for the conventionalist, an improper question (although one could legitimately compare the usefulness of different systems of measurement). But, as we saw in Chapter 1, conventionalism over metric has serious consequences for our view of physical law. How, for example, can we regard the laws of motion as being objectively true if the facts about metric implied in the concepts of velocity and acceleration are merely conventional?

Assuming that space and time are not merely our projections onto the world, what are they? The first place to look will be among the objects of our direct experience, for a failure to perceive space and time contributes to the sense of their unreality. Indisputably, we perceive change. Equally indisputably, we perceive objects as being certain distances from other objects. So perhaps the best strategy for keeping them out there in the world, rather than locked up in the mind, is to identify time with change, and space with the collection of all spatial relations between objects. These are (versions of) *relationism about time*, and *relationism about space*, respectively. We seem, however, to be able to conceive of the idea of time continuing in the absence of change, which obviously would imply that time and change are two different things. What we need is some method of judging the legitimacy of this conception. We may think of a period of time without at the same time thinking of the changes that take place in it, but it does not follow from this that we can think of a period of time that contains *no* changes, or that such a thought is coherent. In Chapter 2, we looked at three influential arguments against the intelligibility, or possibility, of 'empty' time. The strongest of these appealed to the idea that empty time would be causally inert, implying that we would never have any reason to posit its existence.

The existence of spatial vacua is less contentious, but then it does not immediately defeat the view of space as a collection of spatial relations between objects, for these relations are not incompatible with regions of empty space. The difficulty for the relationist, however, is to explain apparent reference to unoccupied spatial points in apparently true statements, for if we can refer to unoccupied parts of space, it seems that these places are objects in their own right, not obviously reducible to facts about things in space. The relationist may be able to dodge this objection, at least temporarily, by distinguishing between truths about the physical world, and abstract geometrical truths. Apparent reference to facts involving unoccupied spatial points could be construed in terms of the latter. These issues were the topic of Chapter 3, in which we also looked at arguments in favour of absolute motion, an idea that implies the existence of space as existing independently of objects.

Part of what makes attempts to reduce space and time to features of objects and events tempting is the conviction that, considered as the *absolutist* considers them, namely as objects in their own right, they would be entirely featureless. In a completely empty and so changeless universe, nothing would distinguish one place or one moment from any other. And such a featureless medium would really explain nothing of what we observe, but would be merely a theoretical abstraction. But would an independently existing space and time really be featureless? In Chapter 4 we looked at the implications the discovery of consistent non-Euclidean geometries had for our understanding of space. First, it undermined the distinction between physical truths and geometrical truths, and so strengthened the argument for the existence of unoccupied spatial points. Secondly, the fact that space has a certain *shape* (curvature, dimensionality, presence or absence of boundaries) can make a real difference to how things can move in space, and so suggests that space can be a *cause*, and not just an impotent medium. The existence of space may be explanatory in another way, too. The spatial properties of certain asymmetric objects such as hands appear to depend on some global property of space itself.

We are pulled in two directions. On the one hand, we feel uncomfortable with the idea of space and time existing in the absence of any concrete objects or events—a discomfort that is particularly strong when we try to imagine time going on in a completely empty universe. On the other hand, treating them just as abstract ways of talking of objects does not do justice to everything we want to say about space and time. Is there any way of resolving this tension? One compromise we canvassed was to treat space as nothing other than the fields of force around and between objects. These would not exist in the absence of any objects, but on the other hand they are something other than those objects, and can exhibit a certain shape that explains the behaviour of objects moving about in apparently empty space. Similarly, we do not have to treat time as wholly reducible to *changes*. Collections of states of affairs, some of them perhaps unchanging states of affairs, would provide alternative building-blocks for time. Combining these two, we have a picture of space and time as an ordered series of states of affairs concerning the properties of and relations between, concrete objects and their fields of force.

The extent to which we think of space and time as independent of their contents will affect our view of their *boundedness* (or unboundedness). Did time have a beginning? Will it have an end? Is there an edge to space? Evidence for a 'Big Bang', in which our universe had its beginning, is, we suggested in Chapter 5, at best very equivocal evidence for a beginning to time. First, the hypothesis of the Big Bang does not necessarily rule out a preceding 'Big Crunch', in which a previously existing universe collapsed. Secondly, to identify the beginning of the universe with the beginning of time is tacitly to make some contentious conceptual presuppositions. This is not to say that those presuppositions are unwarranted, merely that they need to be made explicit and to be justified. What if we do not identify the beginning of the universe with the beginning of time? Then we invite the picture of aeons of empty time preceding the Big Bang, leaving it inexplicable why the Big Bang occurred just when it did, and not earlier or later. Indeed, the Big Bang itself would appear to be an uncaused event. But causal anomalies are also implied in other accounts of time and the universe: if the universe had no beginning in time, but extends infinitely into the past, what then explains its existence? And what if time is *cyclic*, and so has neither a beginning nor an end? Does it not then follow that every event, ultimately, causes itself?

The idea of an edge of space is as difficult to conceive as a beginning of time, but perhaps particularly difficult for the absolutist, who cannot explain the idea in terms of the finitude of the physical universe. As an ancient paradox nicely brings home to us, we find it very hard to say just how things will behave at the edge of absolute space. But, equally, the idea of space going on indefinitely also causes intellectual discomfort. Discussion of these problems in Chapter 6 ended with the suggestion that space may be both finite and unbounded, a view that may be less problematic than its temporal counterpart.

Chapters 5 and 6 were concerned with the infinite *extent* of time and space. In Chapter 7, we turned to their infinite *divisibility*. Intuitively, we think that there is no limit to the extent that an interval of time or region of space can be divided. This seemingly innocuous idea led to a plethora of paradoxes, including two of Zeno's famous paradoxes of motion, the Achilles and the Dichotomy. The essential idea on which these two are based is that, if space and time are infinitely divisible, then any moving object will have to achieve an infinite number of things in a finite time: i.e. pass through an infinite number of sub-distances. Treating these problems simply as mathematical conundrums, requiring for their solution only the technical notion of the infinitesimal, does not do justice to their philosophical interest and importance. Two important philosophical solutions to the paradoxes present themselves: *finitism* and *atomism*. The finitist asserts that there is no set of actually existing concrete objects that is infinite. An interval of time or region of space does not therefore actually contain an infinite number of points. There may nevertheless be no natural limit to the process of dividing an interval or region, and it is this that justifies us in talking of space and time as infinitely divisible. The infinite exemplified by space and time is therefore, in Aristotle's terms, only a *potential infinite*. The problem with the positive part of this proposal is that it leaves it mysterious what grounds the fact that the process of dividing has no natural limit. It is not enough to say that there is nothing that prevents us dividing further: we naturally want to know what it is that *enables* us to go on dividing, and this, surely, is something to do with the structure of space and time. The doctrine of the potential infinite seems an exhortation just to be silent on this structure. Atomism (which is compatible with the negative part of finitism) is not so silent: it asserts that there are spatial and temporal minima, of non-zero magnitude, which represent the limit of any division. This theory has the merit of solving a range of paradoxes: Zeno's Achilles, Dichotomy, and Parts and Wholes paradoxes, Aristotle's conundrum concerning the first and last moment of motion, and Democritus' paradox of the cone. Admittedly, it involves a revision in our ordinary conception of change, and requires us to adopt a non-Euclidean geometry, but we were unable to detect any contradiction in the idea.

So far in our investigation, we had been concerned with problems common to both space and time. But from Chapter 8 onwards, we turned to features that arguably distinguish time from space: the passage, and direction, of time. Although both these features (and perhaps they are not distinct) are deeply familiar ones, pervading our experience as they do, they are not particularly easy to define. The passage of time is often represented in metaphorical terms, typically in terms of a river. The problem with these metaphors is that they typically have time built into them, and so already presuppose a grasp of what the passage of time amounts to. Two ideas are particularly important in articulating the notion: the first is of the changing pastness, presentness, and futurity of events; the second is of events coming into existence and so adding to the total stock of reality. Much of our thought about the passage of time has been dominated by McTaggart's important distinction between two ways of ordering events in time: as an *A-series*—which orders them according to whether they are past, present, or future—and as an *B-series*—which orders them in terms of earlier and later. The key question here is this: given that the facts underlying the two orderings cannot be different, which determines which? Is it the A-series positions that determine the B-series positions, or the other way around? The natural answer is that it is the A-series positions that determine the B-series positions, but this leads straight to McTaggart's famous paradox, which attempts to show that the notion of a real A-series is self-contradictory. This obliged McTaggart himself to deny the reality of time.

Two strategies for coping with the paradox were outlined in Chapter 8: one was *presentism*, the view that only what is present is real, the other was the *B-theory of time*, which regards the B-series as more fundamental than the A-series. Presentism may perhaps articulate our intuitive conception of time, as we naturally regard the past as no longer real and the future as not yet real. It nevertheless faces some formidable difficulties. First, it is by no means clear that it can explain how statements about the past can be true or false. One mechanism presentists might appeal to concerns the causal traces the past leaves on the present: it is these present causal traces, they could argue, that make statements about the past true. But what, in presentist terms, does it *mean* for there to be causal relations between past and present times? Secondly, presentism creates difficulties for our understanding of motion. In one reconstruction of Zeno's Arrow paradox, the topic of Chapter 9, the following problem arose: the presentist is committed to the idea that a moving object must be conceived as moving *in the present*, but is unable to reconcile this with the fact that motion essentially involves facts about an object's position at times other than the present.

In so far as presentists hold that past truths are determined by present fact, it might seem to follow from their position that the past is alterable, in a manner reminiscent of Orwell's dystopian vision in *Nineteen Eighty-Four*. However, one conclusion from our discussion in Chapter 10 was that the presentist is not committed to this dubious position. Indeed, the very notion of the alterability of the past seems to lead inexorably to contradiction. However, there is an important distinction to be made between *altering* facts and *affecting* them. This allows us, both to avoid the fatalist conclusion that, since we cannot change the future, we cannot affect it, and also make sense of time travel, in which present decisions have a causal influence on past events. However, whether time travel is really a coherent notion depends on our understanding of the direction of time (on which more below).

Presentism is one version of the *A-theory of time*, which holds that B-series facts are determined by more fundamental A-series facts. Not all A-theorists are presentists (although those that are not, we suggested above, will have difficulties in escaping from McTaggart's paradox). But whether or not it is combined with presentism, the A-theory faces another problem. If space and time are just the products of our minds, as Kant thought, there are good grounds for thinking them both *unified*: that is, there is just one time and one space, which, if time and space are in the mind, has to be interpreted as meaning that every object of experience is presented as spatially and temporally related to every other. But what if space and time exist independently of our minds? Is there any reason then to think of them as being essentially unified? Here, being unified means that every object and event is really spatially and temporally related to every other. In some contexts, we suggested in Chapter 11, the idea of multiple spaces and time-series (something like the idea of 'parallel universes' in fiction) may have a useful application. Two such contexts are the multiverse hypothesis, entertained by some cosmologists, and the two-slit experiment with light. Now, the idea of multiple spaces, although perhaps surprising, does not, arguably, raise any serious conceptual difficulties. But the idea of multiple time-series is not readily reconcilable with the A-theorist's assertion that B-series facts are determined by A-series facts.

The second strategy we canvassed for dealing with McTaggart's proof of the unreality of time was the *B-theory*. According to this theory, there is no A-series in reality, only a B-series. As a consequence, since the B-series positions of events do not change, there is no passage of time, at least in the way we ordinarily conceive of it. This raises many questions: if there is no A-series, does this entail that statements such as 'The post has just arrived' are *false*? If there is no passage of time, what becomes of our intuitive belief that the future is unreal? And what accounts for the obvious fact that things change, for does, e.g. a cup of tea changing from hot to cold, not require the tea's being hot to have receded into the *past*? Finally, how can the B-theorist explain the direction of time, for surely direction and passage are inextricably linked? These questions raise profound issues, whose surface we have no more than grazed in this discussion, but here is a summary of provisional answers from the B-theorist's perspective:

(i) *A-series truths.* Despite the absence of an A-series in reality, we indisputably have A-series beliefs and give voice to them ('The train has just left', 'The War ended years ago', 'Aunt Jane will be arriving tomorrow'). What makes these beliefs true (or false) are B-series facts. Thus, if the train leaves just prior to 7 a.m., and at 7 a.m. I have the thought that it has just left, then my belief is true. We do not need to appeal to the *pastness* of the train's departure.

(ii) *The reality of the future.* To describe what is past or present on the one hand as *real* and what is future on the other as *unreal,* seems to require a real distinction (and not merely a distinction in thought) between past, present, and future. Since the B-theory denies that there is such a distinction in reality, it follows that, on the B-theory, all times are equally real. (Some B-theorists, it should be pointed out, have attempted to retain something of our ordinary belief in the unreality of the future by relativizing what is unreal to B-series times. So, at any given time, *later* events are unreal. Is this coherent?)

(iii) *Change.* Change, in B-series terms, is just an object's possessing one property at one time, and an incompatible property at a later time. However, for this to be a completely convincing answer, we need to be able to explain how it is that one and the same object can exist first at one time, exhibiting one property, and then at another time, exhibiting a different property. If time did indeed pass, as the A-theorist holds, then objects could, by simply staying in the present, move from one B-series moment to another. But there is no room for such movement in the B-universe.

(iv) *Direction.* In B-series terms, the fact that time has a direction is neither more nor less than the fact that events form a B-series: that is, that they are ordered by the asymmetric earlier than relation.

This last answer requires some expansion. Time, experience tells us, has an intrinsic direction, space not. But what does this actually amount to? Can it really be no more than the asymmetry of the *earlier than* relation? There are asymmetric spatial relations, too, so we need to say a little more than this to explain the difference between time and space. In particular, we need to be able to answer the following questions: why do we experience time as having a direction, from earlier to later? Why does the arrow of time point in the same direction as the causal arrow (from causes to effects), the psychological arrow (from experiences to memories), and the thermodynamic arrow (from order to disorder)? We tackled these questions in Chapter 12, and much of our discussion was taken up with the *causal analysis of time order.* If this analysis is successful, then we have the prospect of being able to solve a number of conundrums concerning the direction of time, as follows: *Q*: Why do causes occur before their effects? *A*: Because 'earlier than' is *defined* in causal terms. *Q*: Why is the 'earlier than' relation asymmetric? *A*: Because the causal relation is asymmetric. *Q*: Why do memories never precede the experiences of which they are memories? *A*: Because the experiences are the causes of those memories. *Q*: Why do we have a sense of the direction of time? *A*: Because we remember the past, perceive only the present (strictly, the very recent past), but never remember or perceive the future. *Q*: And what explains these facts? *A*: Perception and memory are causal processes. Perceiving or remembering the future would entail backwards causation.

However, unless all events are causally related, the causal analysis appears to imply the possibility that time may go in different directions in different parts of the universe.

So, finally, what of the human significance of these issues? Our view of ourselves is intimately bound up with space, time, and causality: we take up space and move about in it, we are affected by change and are the instigators of change, we persist through time. We think of ourselves, in short, as spatial and temporal agents. What, then, if on investigating the matter, we found that space and time could not, on pain of insoluble paradox, be thought real features of the world? This would have a revolutionary effect on what we think ourselves to be. In particular, we would have to reassess the idea of ourselves as physical beings if to be physical is to occupy space. We would have to take seriously the idea that we were unembodied spirits. Or what if instead we concluded only that a particular feature of space and time was unreal, a result of our projecting a feature of our experience onto the world? For example, suppose we decided that the balance of argument was against our common belief that time passes. How would this affect our view of death? For do we not ordinarily see life as an inexorable movement towards extinction (in this world, at least)? If there is no such movement, what does death amount to? What becomes of our belief that our present existence is somehow more real and significant than our past or future existence? Can we continue to see ourselves as free agents if denying the passage of time implies that what we call the future is as fixed as what we call the past? And what becomes of our conception of ourselves as persisting through time being the same person from moment to moment, irrespective of any change we suffer?

Let me dwell a little longer on this last point. A problem for the B-theory, we noted, was to explain how it can be that one and the same object can exist at one time and exist at another time, without it being the case that the object somehow *moves* from one time to another, implying that time itself passes. Addressing this issue head on, let us now, for the first time, introduce the following radical thought: maybe one and the same object *cannot* be at different times. *What we ordinarily think of as the same object, persisting through time, is in fact a succession of different (though very similar) objects, each unchangingly locked into their own time.* Change is then the having of different and incompatible properties by different (but suitably related) objects. Perhaps the best way to imagine this is to think of time as another dimension of space, and treat objects' apparent persistence through time as just extension in this fourth dimension. We imagine, then, a four-dimensional object, which has different parts at different places in these four dimensions. Although we can see the different parts in three of the dimensions as different, we experience different parts in the fourth dimension *as if* they were one and the same thing, moving through time. There is much that is misleading in this description. Time is *not* just a fourth dimension of space, and the B-theorist does not have to say that it is. But this picture nevertheless gives us an intuitive sense of how much we may need to revise our ordinary conception of the persistence of objects through time if the B-theory is correct. (Again, some B-theorists do not accept that we have to give up the idea that one and the same object persists through time. But what alternative account can they offer?) And if we do revise our conception of how ordinary objects persist through time, then we must also revise our conception of our own persistence through time. This may mean giving up the idea that we are the same person from time to time. As Wells's Time Traveller puts it:

For instance, here is a portrait of a man at eight years old, another at fifteen, another at seventeen, another at twenty-three, and so on. All these are evidently sections, as it were, Three-Dimensional representations of his Four-Dimensional being, which is a fixed and unalterable thing.

Apart from these particular ways in which the philosophy of time and space impinge on our self-conceptions, simply contemplating these difficult and abstract issues widens our view of the world. Recalling Roger Bacon's legendary head of brass, we may find the mysteries of space and time unsettling, but there is reason to hope that part of their solution, at least, lies within the compass of human understanding.

> And yet as angels in some brighter dreams
> Call to the soul when man doth sleep;
> So some strange thoughts transcend our wonted themes,
> And into glory peep.
>
> Henry Vaughan, 'They are all gone into the world of light'

Mr Dunne's Dream and Other Problems

1. Mr Dunne's Dream

The following incident is narrated in J. W. Dunne's *An Experiment with Time*, which was first published in 1927. It was the first of many similar experiences, although some were considerably more disturbing than this one.

It occurred in 1898, when I was staying at an hotel in Sussex. I dreamed, one night, that I was having an argument with one of the waiters as to what was the correct time. I asserted that it was half-past four in the afternoon: he maintained that it was half-past four in the middle of the night. With the apparent illogicality peculiar to all dreams, I concluded that my watch must have stopped; and on extracting that instrument from my waistcoat pocket, I saw, looking down on it, that this was precisely the case. It had stopped—with the hands at half-past four. With that I awoke....

I lit a match to see whether the watch had really stopped. To my surprise it was not, as it usually is, by my bedside. I got out of bed, hunted round, and found it lying on the chest of drawers. Sure enough, it *had* stopped, and the hands stood at half-past four.

The solution seemed perfectly obvious. The watch must have stopped during the previous afternoon. I must have noticed this, forgotten it, and remembered it in my dream. Satisfied on that point, I rewound the instrument, but, not knowing the real time, I left the hands as they were.

On coming downstairs next morning, I made straight for the nearest clock, with the object of setting the watch right. For if, as I supposed, it had stopped during the previous afternoon, and had merely been rewound at some unknown hour of the night, it was likely to be out by several hours.

To my absolute amazement I found that the hands had lost only some two or three minutes—*about the amount of time which had elapsed between my waking from the dream and rewinding the watch.* (Dunne 1934, 41–3)

Some questions to consider: What is the significance of the italicized remark? Assuming that the whole incident is pure coincidence, exactly how probable was it that the position of the watch in the dream should correspond with the actual position of the hands? On the suggestion that there might be some process by which he 'saw' the actual position of the hands whilst asleep, Dunne adds the following comment:

Even supposing that there existed unknown rays which could effect that sort of penetration, and then produce vision—which I did not believe—the watch had been lying at a level above that of my eyes. What sort of rays could these be which bent round corners? (Ibid. 43)

What other explanation of the experience could there be? Could it have any consequences for any of the following: the reality of the future, the A-theory of time, the causal theory of time order?

2. Measuring Time

St Augustine (AD 354–430) was a theologian and Bishop of Hippo in Africa. His autobiographical *Confessions* contains a long chapter on the nature of time. One of the problems Augustine considers is whether, and how, we measure time passing:

Does my soul speak the truth to you when I say that I can measure time? I do indeed measure it, but I do not know what I measure. By means of time I measure the movement of bodies. Does this not mean that I measure time itself? Could I measure the movement of a body, that is, measure how long the movement lasted and how long the body took to move between two points, unless I measured the time in which it moved? (Augustine 398, 274)

However, there is a difficulty:

> We cannot measure it if it is not yet in being, or if it is no longer in being, or if it has no duration, or if it has no beginning and no end. Therefore we measure neither the future nor the past nor the present nor time that is passing. Yet we do measure time. (Ibid. 275)

To put it slightly differently: when we measure time, we appear to measure something that both exists and has duration. But neither the past nor future exist, and the present has no duration. Yet there is nothing to time except past, present, and future. What then are we measuring? (Augustine's own view was that time is all in the mind. Does this solve the difficulty?)

3. A Directionless World?

Try to think of a world in which absolutely nothing is happening, and in which nothing ever happened or ever will happen. If time exists in such a world, what would give it its direction? Or would it have no direction in that world? *Can* time exist and not have a direction?

4. Alien numbers

Imagine a world in which $2 + 2 = 5$. Could there be different but consistent arithmetics in just the same way as there are different but consistent geometries?

5. Representing the Fourth Dimension

In H. G. Wells's *The Time Machine,* the Time Traveller is explaining to his incredulous friends the reasoning behind his conviction that time travel is possible. He begins by discussing the nature of space:

Space, as our mathematicians have it, is spoken of as having three dimensions, which one may call Length, Breadth, and Thickness, and is always definable by reference to three planes, each at right-angles to the others. But some philosophical people have been asking why *three* dimensions particularly—why not another direction at right-angles to the other three?—and have even tried to construct a Four-Dimensional geometry. Professor Simon Newcomb was expounding this to the New York Mathematical Society only a month or so ago. You know how on a flat surface, which has only two dimensions, we can represent a figure of a three-dimensional solid, and similarly they think that by models of three dimensions they could represent one of four—if they could master the perspective of the thing. See?

Is there something questionable about the idea that we could represent four dimensions by a three-dimensional model in the same or similar way that we represent three-dimensions in a two-dimensional diagram?

6. *Archytas at Time's Beginning*

'Whatever event we take as the first event, it is always logically possible for there to have been an event before it. But that possibility requires time before the first event, so time itself has no beginning.' Is there anything wrong with this line of reasoning?

7. *Democritus' Trick*

Recall Democritus' cone paradox: take a cone, slice it in half horizontally and compare the areas of the surfaces you have just exposed. If they are equal, then it was not a cone at all but a cylinder; if unequal, then the cone was not smoothly sloping, but stepped. What, if anything, is wrong with this suggestion: 'The areas of the exposed surfaces are the same because they are the *same* surface; it is only because two different descriptions of the surface are given that we think there are two surfaces'?

8. *The Plattner Case*

In another of H. G. Wells's stories, *The Plattner Story*, a schoolmaster, Gottfried Plattner, is conducting an experiment in the school chemistry laboratory. There is an explosion, and Plattner disappears for several days. On his return, he is strangely reversed: his heart is now on his right-hand side, features originally on one side of his face have shifted to the other side, and he has difficulty in writing from left to right. What (if you have not read the story) is your hypothesis about what happened to the unfortunate Plattner?

9. *The Elusive Present*

Suppose the following to be both true:

(i) Time is infinitely divisible, so that between any two moments there is always a third.

(ii) There is a last moment of the past—in other words, a past moment later than which there are no other moments that are also past.

Which moment is present?

10. *Suspicious Success*

In the urn case described at the beginning of Chapter 11, we appealed to the principle that, other things being equal, a hypothesis that makes our observations less improbable is to be preferred to a hypothesis that makes things more improbable. But if we win the jackpot in a lottery in which millions of people took part, we are not inclined to assume that the lottery must have been somehow fixed in our favour, although that would make the outcome less improbable than if the choice of winner had been genuinely random. So does this show that the principle is a flawed one?

11. An Infinite Regress

First-order change is change in the ordinary properties of things, as when a cup of tea cools down. Call such changes simply 'events'. Second-order change, if it exists, is change in those events themselves as they cease to be present and become increasingly past. What would third-order change be? Does first-order change entail the existence of second-order change? Does second-order change entail third-order change, and so on?

12. Does Time Fly in Two Dimensions?

If time really is passing, it is a form of movement (for past events keep receding further from us). But movement is something we measure against *time*. In which case, we should be able to say how much time 5 minutes takes to pass. To which the only answer seems to be that the passing of 5 minutes takes 5 minutes: is this disappointing triviality all that the profound mystery of the passage of time amounts to? There must be more to it than this. If we want a non-trivial answer to the question 'How long does 5 minutes take?' we have, it seems, to appeal to *another dimension,* against which movement in time is measured. This second dimension cannot be a dimension of space, for the passing of any length of time need not involve movement across space at all. So we are led to the idea of a second temporal dimension. Since we need to be able to refer to the two dimensions separately, we will call the first, more familiar, dimension *time*$_1$ and the second dimension *time*$_2$.

Our picture of the passage of time now looks like this (Figure 31). The y-axis represents locations in time$_1$ (the time in which we usually measure changes in the world), such as 13.00 hours. The x-axis represents locations in time$_2$ (the time against which we measure changes in time$_1$). The variably inclined line plots the passage of the *present* from a particular event, located at the origin of the graph

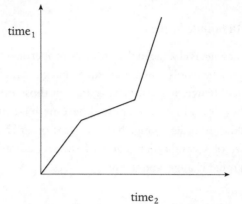

Fig. 31. Two-dimensional time

(the intersection of the two axes). So, e.g. when it is now 13.00 hours in time$_1$, it is now 10.45 *schmours* (say) in time$_2$. We can see, then, that after a certain period, the passage of time slowed down (i.e. the same period in time$_1$ was now taking up more time$_2$ than previously) before speeding up again.

Does this way of looking at things actually work? Can we explain time's flow by appealing to another dimension? Could two-dimensional time offer a way of explaining how time could go *backwards*?

FURTHER READING

1. *The Measure of All Things*

A gripping account of the pursuit of accuracy in timekeeping, centred on John Harrison's long wrestle, between 1727 and 1773, with the problem of devising a timepiece that would remain perfectly accurate at sea (thus allowing sailors to calculate their longitude) is Sobel (1996). O. K. Bouwsma's story is reprinted in Westphal and Levenson (1993). A classic defence of conventionalism about metric is Reichenbach (1958). A very helpful introduction to the debate between conventionalism and objectivism, which clearly sets out the different responses to the threat of inaccessible facts and proposes a novel solution is Newton-Smith (1980, ch. VII). The connection between the debate over metric and the laws of motion is discussed in Van Fraassen (1980, ch. III.2).

2. *Change*

Aristotle's treatment of the relationship between time and change, and his definition of time, can be found in his *Physics*, bk. IV: Hussey (1983). For a helpful discussion of Aristotle's account, see Lear (1982). For Leibniz's views on time, see his correspondence with Samuel Clarke: Alexander (1956), and his *New Essays on Human Understanding*: Remnant and Bennett (1981). A well-known and fascinating thought-experiment designed to show the limitations of verificationist attacks on the notion of temporal vacua is Shoemaker (1968). For a discussion and extension of this thought-experiment, see Newton-Smith (1980, ch. II). Another argument for the possibility of time without change is discussed in Lucas (1973). A detailed discussion of relationism about time is Hooker (1971). Butterfield (1984) is salutary and required reading for anyone who is tempted to analyse times in terms of possible events.

3. A Box with No Sides?

Descartes's letters to Princess Elizabeth and Henry More can be found in Cottingham, Stoothoff, Murdoch, and Kenny (1991). An excellent historical survey of theories of space is Jammer (1969); ch. 2 explores the theological dimensions of the debate over absolute space. For Aristotle's arguments concerning the void see the *Physics*, bk. III (Hussey 1983). For discussion of Aristotle's and other ancient views on the void, see Sorabji (1988). A collection of classic readings on space is Huggett (1999); see especially ch. 4 for extracts from Aristotle and commentary, and ch. 7 for extracts from Newton's *Principia* (including the famous bucket experiment) and commentary. For Leibniz's views on the relationship between space and the objects within it, see again his correspondence with Samuel Clark: Alexander (1956); extracts and commentary in Huggett (1999, ch. 8). The relationist–absolutist debate, and its relationship to the debate over absolute motion, are discussed in van Fraassen (1980, ch. IV.1), and in greater detail by Nerlich (1994a) and Dainton (2001). See also Hooker (1971). A very detailed, but difficult, examination of the debate is Earman (1989).

4. Curves and Dimensions

For an accessible introduction to non- Euclidean geometry, see Sawyer (1955, ch. 6). For the history of geometry see Boyer (1968), especially chs. VII and XXIV. See also van Fraassen (1985, ch. IV.2). The significance of non-Euclidean geometry for the debate between the absolutists and relationists is very clearly articulated in Nerlich (1991) and Dainton (2001). Kant's argument concerning incongruent counterparts is presented in Kant (1768); an extract is reprinted in Huggett (1999, ch. 11), which also provides commentary. See also Earman (1989, ch. 7) and Walker (1978, ch. IV), which provide useful exegeses of Kant's text, and Nerlich (1994a, ch. 2), a particularly important discussion that reconstructs the argument in terms of chirality (he uses the term *enantiomorphism)*, and on which my account is based. A very entertaining discussion of chirality and its philosophical and scientific interest is Gardner (1982).

5. *The Beginning and End of Time*

For a brief introduction to Big Bang cosmology, see Hawking (1986). Ancient arguments concerning creation and the beginning of time are discussed in Sorabji (1983, pt. III). Newton-Smith (1980) exploits number analogies in defining the beginning and end of time, discusses a number of arguments against a beginning, including ones put forward by Aristotle and Kant, and also considers the causal anomalies involved in a beginning and end of the universe. Kant's argument concerning infinity is explored in Moore (1990). Craig and Smith (1993) is an extended, and highly dialectical, exploration of the interaction between Big Bang cosmology, philosophy, and theology, including an assessment of Kant's arguments. For some background to Eliot's *Four Quartets,* see Tamplin (1988). Closed, or cyclic time is discussed at some length in Newton-Smith (1980, ch. III). For historical sources for cyclic time and history, see Sorabji (1983, ch. 12).

6. *The Edge of Space*

Archytas' argument, and related arguments, against the edge of space are presented and discussed in Sorabji (1988i, ch. 8). He also discusses arguments concerning extracosmic space. The difficulties the notion of an edge pose for our understanding of motion are pointed out in Nerlich (1994b). For discussion of Kant's First Antimony, see Broad (1978, ch. 5) and Bennett (1974, ch. 7). For readings on Aristotle's distinction between the actual/potential infinite distinction, see references under Chapter 7 below. An attempt to make sense of Kant's obscure argument from incongruent counterparts to the ideality of space is made in Broad (1978, ch. 2) and van Cleve (1999). Poincaré's thought experiment is presented in his

(1952); extracts and discussion in Huggett (1999, ch. 13). For a discussion of closed space see Sorabji (1988, ch. 10).

7. Infinity and Paradox

The sources of Zeno's arguments can be found in Kirk, Raven, and Schofield (1983), Barnes (1978,) and Huggett (1999). For discussion, see Huggett (1999), Owen (1957–8), Salmon (1970), Sorabji (1983), and Sainsbury (1988). For Aristotle's actual–potential infinite distinction, and its application to Zeno's paradoxes, see Lear (1988). The source of Thomson's Lamp is Thomson (1954). For discussion, see Sainsbury (1988). See also Clark and Read (1984), which argues against the possibility of completing uncountably many tasks in a finite time. Moore (1990) is a particularly helpful and wide-ranging discussion, covering all the above. It also presents an historical overview on theories of infinity and defends a version of finitism. The transition paradox is discussed at length in Sorabji (1983). Part V of Sorabji also discusses Ancient atomism, on which see also Barnes (1978) and Kirk, Raven, and Schofield (1983), for sources and commentary. Lloyd (1982, ch. 4) shows how ancient atomism developed as a response to the problem of change (the problem being that change appears to involve something coming into being from nothing).

8. Does Time Pass?

Broad's views on time underwent an interesting development. Compare his (1923) discussion, from which the passage quoted is taken, with his encyclopaedia article (1921) and his exhaustive discussion (1938) of McTaggart's proof. McTaggart's proof originally appeared in his (1908), but see also the revised presentation in his (1927). For discussion and partial defence of the proof see Dummett (1960). A more critical discussion, which focuses on the legitimacy, or otherwise, of expressions such as 'is present in the past', is Lowe (1987). A reconstruction in terms of token-reflexives is presented in Mellor (1981) and (1998). An unusual solution to the paradox, in terms of possible worlds, is presented in Bigelow (1991), and criticized in Oaklander (1994). Early statements of presentism are Lukasiewicz (1967) and Prior (1970). Versions of presentism are defended in Bigelow (1996) and Zimmerman (1998). The doctrine, in its various forms, is examined in Dainton (2001, ch. 6). The B-theory is defended in Smart (1980), Mellor (1981, 1998), and Oaklander (1984). The most extensive critiques of the B-theory to date are Smith (1993) and Craig (2000). See also Teichmann (1995). For an important collection of papers on McTaggart's argument, the B-theory, and related issues, see Oaklander and Smith (1994). One issue not discussed here is the extent to which the A-theory, and the associated view of the future is unreal, is compatible with the Special Theory of Relativity. For arguments in favour of the view that they are incompatible, see Putnam (1967), Mellor (1974), and Nerlich (1998). For attempts at reconciliation, see Smith (1993, 1998), Dorato (1995), and Craig (2001). See also Dainton (2001, chs. 16 and 17) for an introduction to Special Relativity and its philosophical consequences.

9. *Change again: Zeno's Arrow*

For sources and commentary on the Arrow, see Lee (1936), Barnes (1982), Kirk, Raven, and Schofield (1983), and Huggett (1999). For discussion, see Ross (1936), Owen (1957–8), Vlastos (1966), Grunbaum (1967), Salmon (1970), and Sorabji (1983). The suggestion that the Arrow is best understood in terms of motion in the present was made by Jonathan Lear, and his discussion is particularly helpful: see Lear (1981, 1988). The second reconstruction presented here is based (with a few details altered) on his account. For discussions of presentism, see the references under Chapter 8 above.

10. *Interfering with History*

A very readable history of the calendar, with an account of the shift from the Julian to the Gregorian calendar, the problems that it was intended to solve, and the consequences, is Duncan (1998). For an introduction to the issues involving the reality/unreality of the future, and its bearing on our status as free agents, see Smith and Oaklander (1995). Discussion of these issues as they arose in ancient times is provided by Hintikka (1973), Sorabji (1980), and Lucas (1989). The last of these defends a version of the 'open future' account. On the reality of the past, see Dummett (1969), who defends an 'anti-realist' account. For a discussion of the philosophical dimensions of the passage from Orwell's *Nineteen Eighty-Four* quoted here, see Wright (1986). On time travel, see Harrison (1971), Lewis (1978) (introduces the external–personal time distinction, and argues that the time-traveller is a free agent in the sense that anyone is a free agent), MacBeath (1982) (explores the causal anomalies of time travel), and Ray (1991, ch. 8) (puts time travel in the context of spacetime physics). For a defence of the coherence of backwards causation, see Dummett (1964), and arguments for its incoherence, see Mellor (1981, 1998). Mellor's argument in his (1981) is criticized in Riggs (1991) and Weir (1988) (who puts the issue in the context of closed, or cyclic, time).

11. *Other Times and Spaces*
An excellent and very readable discussion of the multiverse hypothesis, its theistic rival, and the probabilistic reasoning used to motivate them, is Leslie (1989). Leslie also briefly discusses the two-slit experiment. The branching spaces interpretation of the experiment is closely related to what is often called the 'many worlds' interpretation of quantum mechanics (although there is considerable dispute over the correct interpretation of the interpretation). For a philosophical introduction to quantum mechanics, see Lockwood (1989). The many worlds, or 'many minds', interpretation was the subject of a symposium, the (highly technical) papers being published in the June 1996 issue of the *British Journal for the Philosophy of Science*. Disunified space is the subject of a classic paper by Quinton (1962). In that paper Quinton puts forward a fantasy designed to show that, under certain conditions, we might regard our experi-

ence as providing evidence for disunified space. He rejects the temporal parallel, however, as does Swinburne (1981, ch. 10), which nevertheless considers it in sympathetic detail (see also ch. 2 on disunified space). The temporal parallel is given support in Newton-Smith's (1980) discussion of Quinton's thought experiment. See also Hollis (1967).

12. *The Arrows of Time*

An important recent discussion of the arrow(s) of time, which clearly presents the view that the classic way of articulating the problem of the direction of time is ill-posed, and which defends a novel treatment of causation, is Price (1996). The issue of what kind of reduction is appropriate in reductionist theories of direction is taken up by Sklar (1981), who argues in favour of a theoretical reduction rather than one in terms of meaning. For a discussion of the psychological arrow, see Newton-Smith (1980). The thermodynamic and causal arrows are discussed in Dainton (2001), which also tackles the tricky issue of how the asymmetry of causation is to be explained. The causal analysis of time is subject to lengthy and detailed criticism in Sklar (1974). Tooley (1997) is an intriguing attempt to explain temporal and causal asymmetry in terms of the unreality of the future, but appealing only to B-series facts. The importance of causation in explaining our experience of time, and in particular its direction, is well articulated in Mellor (1981, 1998). On the issue of dimensionality: a two-dimensional model for time's passage is presented in Schlesinger (1982) and criticized by MacBeath (1986). For an ingenious discussion of what might count as evidence for two-dimensional time, see MacBeath (1993).

Concluding Thoughts

The human significance of time, as revealed by the history of Western and Eastern thought about time, is discussed in Fraser (1968, pt. I). On the significance of the A-theory / B-theory debate for our views on death, see Le Poidevin (1996). The reconciliation of human freedom with the B-theory is the subject of Oaklander (1998). The ethical dimension of the debate is explored in Cockburn (1997, 1998). The issue of whether the B-theory requires a revision in our ordinary views of persistence through time has been the subject of much debate. For arguments that it does, see Lowe (1998a, 1998b); that it does not, Mellor (1981, 1998).

BIBLIOGRAPHY

Alexander, H. G. (ed.) (1956) *The Leibniz–Clarke Correspondence*, Manchester: Manchester University Press.

Augustine, St (398) *Confessions*, ed. R.D. Pinecoffin (1961), Harmondsworth: Penguin.

Barnes, Jonathan (1982) *The Presocratic Philosophers*, London: Routledge & Kegan Paul.

Bennett, Jonathan (1974) *Kant's Dialectic*, Cambridge: Cambridge University Press.

Bigelow, John (1991) 'Worlds Enough for Time', *Noûs*, 25, 1–19.

—— (1996) 'Presentism and Properties', *Philosophical Perspectives* 10, 35–52.

Boyer, Carl B. (1968) *A History of Mathematics*, 2nd imp., Princeton: Princeton University Press, 1985.

Brand, M. (1980) 'Simultaneous Causation', in van Inwagen (1980), 137–53.

Broad, C. D. (1921) 'Time', in J. Hastings (ed.), *Encyclopaedia of Religion and Ethics*, Edinburgh, 334–45.

—— (1923) *Scientific Thought*, London: Routledge & Kegan Paul.

—— (1938) *An Examination of McTaggart's Philosophy*, vol. II, pt. I, Cambridge: Cambridge University Press.

—— (1978) *Kant: An Introduction*, ed. C. Lewy, Cambridge: Cambridge University Press.

Butterfield, Jeremy (1984) 'Relationism and Possible Worlds', *British Journal for the Philosophy of Science*, 35, 101–13.

Clark, Peter, and Read, Stephen (1984) 'Hypertasks', *Synthese*, 61, 387–90.

Cockburn, David (1997) *Other Times: Philosophical Perspectives on Past, Present and Future*, Cambridge: Cambridge University Press.

—— (1998) 'Tense and Emotion', in Le Poidevin (1998), 77–91.

Cottingham, John, Stoothoff, Robert, Murdoch, Dugald, and Kenny, Anthony, trans. (1991) *The Philosophical Writings of Descartes, iii: The Correspondence*, Cambridge: Cambridge University Press.

Craig, William Lane (2000) *The Tenseless Theory of Time: A Critical Examination*, Dordrecht: Kluwer.

—— (2001) *Time and the Metaphysics of Relativity*, Dordrecht: Kluwer.

—— and Smith, Quentin (1993) *Theism, Atheism, and Big Bang Cosmology*, Oxford: Clarendon Press.

Dainton, Barry (2001) *Time and Space*, Chesham: Acumen.

Dorato, Mauro (1995) *Time and Reality: Spacetime Physics and the Objectivity of Temporal Becoming*, Bologna: Cooperativa Libraria Universitaria Editrice Bologna.

Dummett, M. A. E. (1960) 'A Defense of McTaggart's Proof of the Unreality of Time', *Philosophical Review*, 69; repr. in Dummett (1978), 351–7.

—— (1964) 'Bringing About the Past', *Philosophical Review*, 73, 338–59; repr. in Le Poidevin and MacBeath (1993), 117–33.

—— (1969) 'The Reality of the Past'; repr. in Dummett (1978), 358–74.

—— (1978) *Truth and Other Enigmas*, London: Duckworth.

Duncan, David Ewing (1998) *The Calendar*, London: Fourth Estate.

Dunne, J. W. (1934) *An Experiment with Time*, 2nd edn., London: Faber & Faber.

Earman, John (1989) *World Enough and Spacetime*, Cambridge, Mass: MIT Press.

Fraser, J. T. (ed.) (1968) *The Voices of Time*, London: Penguin Press.

Gardner, Martin (1982) *The Ambidextrous Universe*, 2nd edn., Harmondsworth: Penguin.

Grünbaum, Adolf (1967) *Modern Science and Zeno's Paradoxes*, Connecticut: Wesleyan University Press.

Harrison, Jonathan (1971) 'Dr Who and the Philosophers', *Proceedings of the Aristotelian Society*, supp. vol. 45, 1–24.

Hawking, Stephen W. (1986) *A Brief History of Time: from the Big Bang to Black Holes*, London: Bantam Press.

Hintikka, Jaakko (1973) *Time and Necessity*, Oxford: Clarendon Press.

Hollis, Martin (1967) 'Time and Spaces', *Mind*, 76, 524–36.

Hooker, Clifford A. (1971) 'The Relational Doctrines of Space and Time', *British journal for the Philosophy of Science*, 22, 97–130.

Huggett, Nick (ed.) (1999) *Space from Zeno to Einstein*, Cambridge: Mass.: MIT Press.

Hussey, Edward (ed.) (1983) *Aristotle's Physics, Books III and IV*, Oxford: Clarendon Press.

Jammer, Max (1969) *Concepts of Space: The History of Theories of Space in Physics*, 2nd edn., Cambridge, Mass.: Harvard University Press.

Kant, Immanuel (1768) 'Concerning the Ultimate Foundation of the Differentiation of Regions in Space', in G. B. Kerford and D. E. Walford, *Kant: Selected Precritical Writings*, (1968), 36–43, Manchester: Manchester University Press.

—— (1787) *Critique of Pure Reason*, 2nd edn., trans. by Norman Kemp Smith, 2nd imp., 1933, London: Macmillan.

Kirk, G. S., Raven, J. E., and Schofield, M. (1983) *The Presocratic Philosophers*, 2nd edn., Cambridge: Cambridge University Press.

Lear, Jonathan (1981) 'A Note on Zeno's Arrow', *Phronesis*, 26, 91–104.

—— (1988) *Aristotle: The Desire to Understand*, Cambridge: Cambridge University Press.

Lee, H. D. P. (1936) *Zeno of Elea*, Cambridge: Cambridge University Press.

Le Poidevin, Robin (1996) *Arguing for Atheism: An Introduction to the Philosophy of Religion*, London: Routledge.

—— (ed.) (1998) *Questions of Time and Tense*, Oxford: Clarendon Press.

—— and MacBeath, Murray (eds.) (1993) *The Philosophy of Time*, Oxford: Oxford University Press.

Leslie, John (1989) *Universes*, London: Routledge.

Lewis, David (1978) 'The Paradoxes of Time Travel', *American Philosophical Quarterly*, 13, 145–52; repr. in Le Poidevin and MacBeath (1993), 134–46.

Lloyd, G. E. R. (1982) *Early Greek Science: Thales to Aristotle*, London: Chatto & Windus.

Lockwood, Michael (1989) *Mind, Brain and the Quantum*, Oxford: Blackwell.

Lowe, E. J. (1987) 'The Indexical Fallacy in McTaggart's Argument for the Unreality of Time', *Mind* 96, 62–70.

—— (1998a) *The Possibility of Metaphysics*, Oxford: Clarendon Press.

—— (1998b) 'Time and Persistence', in Le Poidevin (1998), 43–59.

Lucas, J. R. (1973) *A Treatise on Time and Space*, London: Methuen.

—— (1989) *The Future*, Oxford: Blackwell.

Lukasiewicz, J. (1967) 'Determinism', in Storrs McCall (ed.), *Polish Logic 1920–1939*, Oxford: Oxford University Press.

MacBeath, Murray (1982) 'Who Was Dr Who's Father?', *Synthese*, 51, 397–430.

—— (1986) 'Clipping Time's Wings', *Mind*, 95, 233–7.

—— (1993) 'Time's Square', in Le Poidevin and MacBeath (1993), 183–202.

McCall, Storrs (1994) *A Model of the Universe*, Oxford: Oxford University Press.

McTaggart, J. Mc. E. (1908) 'The Unreality of Time', *Mind*, 17, 457–74.

McTaggart, J. Mc. E. (1927) *The Nature of Existence*, ii, Cambridge: Cambridge University Press; ch. 33 repr. in Le Poidevin and MacBeath (1993), 23–34.

Mellor, D. H. (1974) 'Special Relativity and Present Truth', *Analysis*, 34, 74–8.

—— (1981) *Real Time*, Cambridge: Cambridge University Press.

—— (1998) *Real Time II*, London: Routledge.

Moore, A. W. (1990) *The Infinite*, London: Routledge.

Nerlich, Graham (1991) 'How Euclidean Geometry Has Misled Metaphysics', *Journal of Philosophy* 88, 69–83.

—— (1994a) *The Shape of Space*, 2nd edn., Cambridge: Cambridge University Press.

—— (1994b) *What Spacetime Explains*, Cambridge: Cambridge University Press.

—— (1998) 'Time as Spacetime', in Le Poidevin (1998), 119–34.

Newton-Smith, W. H. (1980) *The Structure of Time*, London: Routledge & Kegan Paul.

Oaklander, Nathan (1984) *Temporal Relations and Temporal Becoming: A Defense of a Russellian Theory of Time*, Lanham: University Press of America.

—— (1994) 'Bigelow, Possible Worlds and the Passage of Time', *Analysis*, 54, 159–66.

—— (1998) 'Freedom and the New Theory of Time', in Le Poidevin (1998), 185–205.

—— and Smith, Quentin (eds.) (1994) *The New Theory of Time*, New Haven: Yale University Press.

Owen, G. E. L. (1957–8) 'Zeno and the Mathematicians', *Proceedings of the Aristotelian Society* 58, 199–222; repr. in Owen (1986), 45–61.

—— (1976) 'Aristotle on Time', in P. Machamer and R. Turnbull (eds.), *Motion and Time, Space and Matter*, Columbus: Ohio State University Press, 3–27; repr. in Owen (1986), 295–314.

—— (1986) *Logic, Science and Dialectic*, ed. Martha Nussbaum, London: Duckworth.

Poincaré, Henri (1952) *Science and Hypothesis*, New York: Dover Publications, Inc.

Price, Huw (1996) *Time's Arrow and Archimedes' Point*, Oxford: Oxford University Press.

Prior, A. N. (1970) 'The Notion of the Present', *Studium Generale* 23, 245–8.

Putnam, Hilary (1967) 'Time and Physical Geometry', in H. Putnam, *Mathematics, Matter and Method*, Cambridge: Cambridge University Press, 1975, 198–205.

Quinton, Anthony (1962) 'Spaces and Times', *Philosophy*, 37, 130–47; repr. in Le Poidevin and MacBeath (1993), 203–20.

Ray, Christopher (1991) *Time, Space and Philosophy*, London: Routledge.

Reichenbach, Hans (1958) *The Philosophy of Space and Time*, London: Dover.

Remnant, Peter, and Bennett, Jonathan (eds.) (1981) *Leibniz's New Essays on Human Understanding*, Cambridge: Cambridge University Press.

Riggs, Peter J. (1991) 'A Critique of Mellor's Argument against "Backwards" Causation', *British Journal for the Philosophy of Science*, 42, 75–86.

Ross, W. D. (1936) *Aristotle's Physics*, Oxford: Clarendon Press.

Russell, Bertrand (1903) *The Principles of Mathematics*, Cambridge: Cambridge University Press.

Sacks, Oliver (1986) *The Man Who Mistook His Wife for a Hat*, London: Picador.

Sainsbury, R. M. (1988) *Paradoxes*, Cambridge: Cambridge University Press.

Salmon, W. C. (ed.) (1970) *Zeno's Paradoxes*, Indianapolis: Bobbs-Merrill.

Sawyer, W. W. (1955) *Prelude to Mathematics*, Harmondsworth: Penguin.

Schlesinger, George (1982) 'How Time Flies', *Mind*, 91, 501–23.

Shoemaker, Sydney (1969) 'Time Without Change', *Journal of Philosophy*, 66, 363–81; repr. in Le Poidevin and MacBeath (1993), 63-79.

Sklar, Lawrence (1974) *Space, Time and Spacetime*, Berkeley, Calif.: University of California Press.

—— (1981) 'Up and Down, Left and Right, Past and Future', *Noûs*, 15, 111–29; repr. in Le Poidevin and MacBeath (1993), 99–133.

Smart, J. J. C. (1980) 'Time and Becoming', in Van Inwagen (1980), 3–15.

Smith, Quentin (1993) *Language and Time*, New York: Oxford University Press.

—— (1998) 'Absolute Simultaneity and the Infinity of Time', in Le Poidevin (1998), 135–83.

—— and Oaklander, Nathan (1995) *Time, Change and Freedom*, London: Routledge.

Sobel, Dava (1996) *Longitude*, London: Fourth Estate.

Sorabji, Richard (1980) *Necessity, Cause and Blame*, London: Duckworth.

—— (1983) *Time, Creation and the Continuum*, London: Duckworth.

—— (1988) *Matter, Space and Motion*, London: Duckworth.

Swinburne, Richard (1981) *Space and Time*, 2nd edn., London: Macmillan.

Tamplin, Ronald (1988) *A Preface to T. S. Eliot*, London: Longman.

Teichmann, Roger (1995) *The Concept of Time*, London: Macmillan.

Tooley, Michael (1997) *Time, Tense, and Causation*, Oxford: Clarendon Press.

Thomson, James (1954) 'Tasks and Super-Tasks', *Analysis* 15, 1–13.

Van Cleve, James (1999) *Problems from Kant*, New York: Oxford University Press.

Van Fraassen, Bas C. (1985) *An Introduction to the Philosophy of Time and Space*, 2nd edn., New York: Columbia University Press.

Van Inwagen, Peter (ed.) (1980) *Time and Cause: Essays for Richard Taylor*, Dordrecht: D. Reidel.

—— and Zimmerman, D. (eds.) (1998) *Metaphysics: the Big Questions*, Oxford: Blackwell.

Vlastos, G. (1966) 'A Note on Zeno's Arrow', *Phronesis*, 11, 3–18.

Walker, Ralph (1978) *Kant*, London: Routledge & Kegan Paul.

Weir, Susan (1988) 'Closed Time and Causal Loops: A Defence against Mellor', *Analysis*, 48, 203–9.

Westphal, Jonathan and Levenson, Carl (eds.) (1993) *Time*, Hackett Publishing Company.

Wright, Crispin (1980) 'Realism, Truth-Value Links, Other Minds and the Past', *Ratio*; repr. in Wright (1993, 85–106).

—— (1986) 'Anti-Realism, Timeless Truth and Nineteen Eighty-Four', in Wright (1993), 176–203.

—— (1993) *Realism, Meaning and Truth*, 2nd edn., Oxford: Blackwell.

Zimmerman, Dean (1998) 'Presentism and Temporary Intrinsics', in Van Inwagen and Zimmerman (1998), 206–19.

INDEX

A-series 128–35, 140–3, 144–6,
        197–201, 229–30, 241–3
    see also B-series
A-theory 159, 162, 242–3
    see also presentism
A-universe 140, 142, 159, 162
Abbott, E. A. 56
absolutism:
    about space 37–50, 57–61, 63–6,
        93–5, 99, 161, 236–8
    about time 27–8, 49–50, 237–8
    see also relationism
acceleration 33, 46–50, 58, 65
    argument from absolute 48–50, 65
Achilles paradox 101–2, 104–5, 153,
        154
affecting/changing distinction
        169–70
Alexander of Aphrodisias 91
Antinomies of Pure Reason 80
    First Antinomy 80–3, 92–4
Archytas 89–91, 93, 95, 99
argument:
    from absolute acceleration
        48–50, 65
    Aristotle's, against beginning of
        time 77–8
    Arrow 149–62

from chirality 68–70
against extracosmic space 42–3
from handedness 64–6
Kant's, against spatial infinitude
    of world 95–7
Kant's, for spatial infinitude of
    world 92–4
McTaggart's 131–5, 140–1, 143,
    159, 162, 168, 182, 235, 241–3
measure 22–4, 37
simultaneity of causation 226–8
sufficient reason 26–8, 37, 74–5
from uncompletability 81–3
    see also paradox
Aristotle:
    on the beginning and end of
        time 77–8
    on change and time 14–18
    on infinity 96–7, 106–7, 109,
        110–11, 119, 240
    on reality of space and time 235
    on the void 31–6
    on Zeno 102, 151–4
arithmetic 40–1
Arrow paradox 102, 149–62, 241
asymmetry:
    logical 204–5, 233, 244–5
    spatial 63–4, 94, 238

asymmetry (*cont.*):
   temporal 206–7, 233, 244–5, *see also* time, direction of
atomism:
   of matter 32, 34–5
   of space and time 119–21, 152–4, 239–40
Auden, W. H. 13
Augustine, St, of Hippo 75–6, 235

B-series 128–35, 140–3, 198–201, 241, 243–4
B-theory 140–4, 159, 168–70, 241, 243–7
B-universe 140–3, 159, 169, 197
   see also B-theory
betweenness 215–17, 225–6
Big Bang 75–6, 187, 238–9
Big Crunch 75–6, 238
Bolyai, J. 54
Bouwsma, O. K. 4
Bradley, F. H. 235
Broad, C. D. 125–7, 129–30
bucket experiment 47–50

calendar 164–6
causal arrow of time 206–7, 212–13, 217–33, 244–5
causation 79, 85, 87–8, 138–40, 170, 195
   backwards 181–3
   properties of 87–8
   simultaneous 226–8
   see also causal arrow of time
change 13–28
   and A-series 128, 130–3, 143
   B-theory of 143, 244
   first-order 16–17

in objects and facts 169–70
   second-order 16–17
   time and 15–28, 128–33, 77, 143, 244
   see also vacua, temporal
chirality 61, 63–4, 68–70
   argument from 68–70
   defined 63
Clarke, S. 25
clocks 2–9
Conrad, J. 1–2
continuity 114–15, 118–19
   see also divisibility; infinity
conventionalism about metric 6–8, 10–12, 23–4, 60, 236

death 245
Democritus:
   atomism 32, 136–7
   cone paradox 115–19
density 112
   see also divisibility
Descartes, R. 30–1, 137
Dickens, C. 13–14, 136
Dichotomy paradox 102–3, 104–5, 153, 154
dimensionality:
   defined 68
   of space 36, 56–7, 67–71, 190, 194
   of time 36, 233
direction of time 86, 202–6, 229–33
discreteness 114–15
   see also atomism
divisibility:
   of space 103–6, 110–11, 153, 239–40
   of time 151–4, 239–40
doubling in size 59–60

earlier than relation, properties of 204–5
  see also time order, analyses of
Einstein, A. 191
Eliot, T. S. 83–4
Elizabeth of Bohemia 30–1
entropy 208–13, 217–18, 219
Euclid 53–7
  see also geometry
Euler, L. 10
experience argument 18–22, 36
extracosmic space, argument against 42–3

facts 169–70
fine tuning 187–9, 201
finitism 106–7, 110, 153, 239–40
first cause argument 79
forces 33, 46–50, 60, 61, 65, 70, 91, 111, 238
fourth dimension 67–71, 194, 232, 246–7
Fox Talbot, W. 148
freedom 167–70, 245–6
future:
  affecting 169–70
  reality of 126, 134, 168, 243–4

Galilei, G. 32, 35, 44
Gauss, C. 54, 55
geometry 40–1, 53–61, 99, 120–1
  non-Euclidean 54–61, 121, 237
globes experiment 46–50
God 31, 79, 188, 212
gravitation 33, 70
Greenwich Observatory 1–2
Gregorian calendar 164–6
Guericke, O. von 36

handedness 62–9, 94–5
  argument from 64–6
  defined 64
Hegel, G. W. F. 128
'here' 123, 134–5
Hubble, E. 73–4

incongruent counterparts 63–4, 66, 93, 94–5
infinite 43, 81–3, 92–9, 234
  actual versus potential 96–7, 106–7, 109, 153, 240
  divisibility 102, 103–19, 152–4, 234
infinitesimals 104
intrinsic properties 158
instants 150–7

Johnson, S. 165
Julian calendar 164–6

Kant, I.:
  on beginning of time 80–3
  on handedness and space 62–6, 69
  at Königsberg 62, 79–80
  on reality of space and time 34, 85, 88, 235
  on spatial infinitude of the world 92–7
  on uniqueness of space 193
Korsakov's syndrome 213–14

laws 189–90
  see also motion, Newton's Laws of
Leibniz, G. W.:
  on absolute space and motion 45–6, 50, 59, 60

Leibniz, G. W. (*cont.*):
  on the beginning of the world
    25–6
  correspondence with Clarke 25
  on genealogical analogy of
    space 38–9
  on sufficient reason 25–6, 42
  on time without change 22–4
Lewis, D. 174
light 74, 99, 191–2
Lobachevsky, N. 54
Locke, J. 137, 213

McTaggart, J. McT. E.:
  A-series/B-series distinction, *see*
    A-series; B-series
  at Cambridge 127–8, 146
  proof of unreality of time 131–5,
    140–1, 143, 159, 162, 168, 182,
    235, 241–3
Marey, E. 149
Mary, Queen of Scots 83–4
measure argument 22–4, 37
memory 213–14
  see also psychological arrow of
    time
metaphysical necessity 109–10
metric 6–8, 10–12, 22–4, 60, 236
modality 109–10, 220–1
moments 111–15, 149–63
moncupator 19–20
monism 102
Moore, G. E. 127–8, 146
More, H. 31
motion:
  absolute versus relative 44–50
  dynamic account of 158–9,
    161–2

first moment of, *see* Transition
  paradox
and geometry 58–9
Newton's Laws of 8–12, 60, 70,
  91
perception of 149
static account of 155–6, 157–8,
  160–2
unreality of, *see* Zeno of Elea
see also acceleration
multiverse 188–90
Muybridge, E. 148–9, 155

Newton, I. 46–7, 50
  see also motion, Newton's Laws
    of; bucket experiment; globes
    experiment
'now' 123–4, 143–6, 156–8
  see also time, passage of
numbers 96

objectivism about metric 6–8,
  10–12, 23–4, 236
omnipresence, divine 31
Orwell, G. 170–3, 242

paradox:
  Achilles 101–2, 104–5, 153, 154
  Archytas' 89–91, 93, 95, 99
  Arrow 102, 149–62, 241
  Democritus' cone 115–19
  Dichotomy 102–3, 104–5, 153,
    154
  First Antinomy 80–3, 92–4
  McTaggart's, *see* McTaggart
  Parts and Wholes 103–4
  Thomson's Lamp 107–10
  of time travel 174–81

Transition 111–15
of the void 32–3
Parmenides 24, 102, 235
Parts and Wholes paradox 103–4
past:
alterability of 167–80
reality of 135–40, 170–4, 177–8
persistence 246–7
personal/external time distinction
174–6
photography 148–9
Plato 15, 89, 102
Poincaré, H. 59, 98–9
possibility 39–40, 220–221, 226–7
present 77–8, 143–6, 154, 156–63
see also A-series; presentism
presentism 136–40, 159–63, 170–4,
177–8, 182, 199, 241–2
probability 185–9, 209–11, 213
psychological arrow of time
206–7, 211–18, 219, 231–3, 244–5
Pythagoras' theorem 120–1

Railway Time 2
red shift phenomenon 74
reflexivity 87
relationism:
about space 37–50, 58–61, 64–6,
93–5, 161, 236–8
about time 27–8, 49–50, 236–8
Riemann, G. 55
Russell, B. 127–8, 146, 155
Rutherford, E. 146

Saccheri, G. 54
Sacks, O. 213–14
Simplicius 150
simultaneity 221–5, 226–8

see also B-series
soul 30–1
space:
absolute, see absolutism
atoms of 119–21, 152–4, 239–40
branching 192–3
closed 99
curved 99
continuity/density of 113, 153,
154
see also divisibility
dimensionality of, see dimensionality of space
divisibility of 103–6, 110–11,
239–40
edge of 89–99, 238
empty 31–40, 61
extracosmic 42–3
as a field of force 61, 70, 111,
238
as a form of intuition 81
genealogical analogy of
38–9
geometry and 40–1, 54–61, 121,
237
multiple 190–5
non-Euclidean 54–61, 121, 237
points in 39–41, 60–1
as possibility of location
39–40
reality of 234–5, 242
relationist theory of, see
relationism
unity of 190–5, 242
spatial B-series 135
straight lines 52–6, 99
sufficient reason, principle of 25–8,
78

sufficient reason argument 26–8,
  37, 74–5
synthetic *a priori* 235

tense 129
  see also A-series
tenseless expressions 129
  see also B-series
thermodynamic arrow of time
  206–13, 219, 244–5
Thermodynamics, Second Law of
  207–13
Thomson, J. 107–8
Thomson's Lamp 107–10
time:
  absolutist theory of, *see* abso-
    lutism
  atoms of 119–21, 152–4,
    239–40
  backwards 202–3
  beginning of 75–88, 93
  branching 200
  and causality 195
  and change 15–28, 77; *see also*
    vacua, temporal
  cyclic 84–8, 239
  continuity/density of 112–15,
    152–4
  dimensionality of, *see under*
    dimensionality
  direction of 86, 202–26,
    229–33
  discrete 119–21, 152–4
  divisibility of 239–40
  empty 17–28, 75–6, 236–7
  existence and 177–8, 197–200
  experience of 4–5, 17–19, 148–9,
    231–3, 244–5

external 175–6
  as a form of intuition 81
  human significance of 245–7
  metaphors of 125–7
  metric of, *see* metric
  moments in 111–15, 149–63
  multiple 195–201, 242–3
  order, *see* time order,
    analyses of
  passage of 86–7, 122–46, 229–30,
    232–3, 240, 245–6; *see also*
    A-series
  personal 174–6
  rate of flow of 125–6
  reality of 128–35, 234–6
  relationist theory of *see*
    relationism
  travel in 174–82
  unity of 195–201, 242–3
time order, analyses of:
  causal 218–33, 244–5; *see also*
    causal arrow of time
  psychological 214–18; *see also*
    psychological arrow of
    time
  thermodynamic 210–13; *see also*
    thermodynamic arrow of
    time
time travel 174–82
Torricelli, E. 35–6
Transition paradox 111–15
transitivity 87
two slit experiment 191–3, 201

uncompletability, argument from
  81–3
unoccupied points 39–41, 60–1
  see also vacua, spatial

vacua:
   spatial 31–40, 61
   temporal 17–28, 36–7, 59, 75–6, 80
verificationism 7, 18, 19–20, 23
void, *see* vacua, spatial

Wells, H. G. 232–3, 246–7
Wittgenstein, L. 20–1, 82

Young, T. 191

Zeno of Elea 102–7, 115, 148–59, 235, 241
   Achilles 101–2, 104–5, 153, 154
   Arrow 102, 149–62, 241
   Dichotomy 102–3, 104–5, 153, 154
   Parts and Wholes 103–4